Neotropical Primates

Lynx Illustrated Checklist of Neotropical Primates

Anthony B. Rylands
Russell A. Mittermeier
Jessica W. Lynch
Leandro Jerusalinsky
Karen B. Strier
Liliana Cortés-Ortiz
Andrés Link
Stella de la Torre
Fabiano R. de Melo
Gustavo R. Canale
Jean P. Boubli
Fanny M. Cornejo
Wes Sechrest

First Edition: July 2024

© Lynx Nature Books
 Lynx Nature Books®: Alada Books, S.L.

Based on texts of the *Handbook of the Mammals of the World* (HMW) series, with modifications and updates.

Illustrations by Lluís Sogorb, Ilian Velikov and Francesc Jutglar.

Project co-ordinator: Amy Chernasky
Map production: Daniel Roca
Layout: Daniel Roca
Interior book design: Elena Fonts Circuns

Cover illustration by Faansie Peacock
Southern Muriqui (*Brachyteles arachnoides*)

Printed in Barcelona by Índice Arts Gràfiques.
Legal Deposit: B 14708-2024
ISBN: 978-84-16728-50-3

All rights reserved. No part of this book may be reproduced or transmitted in any form or by any means, electronic or mechanical, including photocopy, recording or any information retrieval system without the prior written permission of Alada Books, S.L.

This book is dedicated to

Alcides Pissinatti

Director of the *Centro de Primatologia do Rio de Janeiro* (CPRJ / INEA), a facility recognized internationally for more than 45 years for its importance in the conservation and the reproduction *ex situ* of Brazilian primates.

A highly skilled veterinary pathologist, Alcides Pissinatti, is one of the great pioneers of Brazilian primatology and one of the most respected Brazilian scientists today. He has long played a major role in the development of Primatology in Brazil and South America. His lifelong career has been devoted to the conservation and breeding of endangered primates, notably the lion tamarins, but many other species as well. Working in the Fundação Estadual de Engenharia do Meio Ambiente (FEEMA), in 1979 he played a key role in the creation of the Rio de Janeiro Primatology Center, founded by Adelmar Faria Coimbra-Filho. Pissinatti has always been a true gentleman, willing to help his colleagues and students in every way possible. It is entirely appropriate that this book be dedicated to him and his extraordinary career.

AUTHORS

Anthony B. Rylands
Deputy Chair, IUCN SSC Primate Specialist Group; and Primate Conservation Director, Re:wild, Austin, Texas, USA.

Russell A. Mittermeier
Chair, IUCN SSC Primate Specialist Group; and Chief Conservation Officer, Re:wild, Austin, Texas, USA.

Jessica W. Lynch
Institute for Society and Genetics and Department of Anthropology, University of California, Los Angeles, California, USA.

Leandro Jerusalinsky
Centro Nacional de Pesquisa e Conservação de Primatas Brasileiros - CPB, Instituto Chico Mendes de Conservação da Biodiversidade - ICMBio, Cabedelo, Paraíba, Brazil.

Karen B. Strier
Department of Anthropology, University of Wisconsin–Madison, Madison, Wisconsin, USA.

Liliana Cortés-Ortiz
Department of Ecology and Evolutionary Biology, University of Michigan, Ann Arbor, Michigan, USA.

Andrés Link
Departamento de Ciencias Biológicas, Universidad de Los Andes, Bogotá, Colombia.

Stella de la Torre
Colegio de Ciencias Biológicas y Ambientales, Universidad San Francisco de Quito, Quito, Ecuador.

Fabiano R. de Melo
Departamento de Engenharia Florestal, Universidade Federal de Viçosa, Viçosa, Minas Gerais, Brazil.

Gustavo R. Canale
Applied Ecology Group, Social, Humanities and Nature Sciences Institute, Sinop, Mato Grosso, Brazil.

Jean P. Boubli
School of Science, Engineering & Environment, University of Salford, Salford, UK.

Fanny M. Cornejo
Yunkawasi, Lima, Peru.

Wes Sechrest
Chief Scientist, CEO and Board Chair, Re:wild, Austin, Texas, USA.

CONTENTS

FOREWORD 11
INTRODUCTION 13
    The Neotropical Region 13
    The Neotropical primates 13
    Diversity 15
    New species 19
    Three mystery monkeys 19
    Quaternary fossil primates 20
    Their habitats 23
    Threats 27
    Conservation status 28
    Primate watching and primate tourism 31
    References 34
USING THE ILLUSTRATED CHECKLIST 39
    Background 39
    Geographical scope 39
    Taxonomic treatment 39
    Changes at the genus level 39
    New species 40
    Taxonomic re-arrangements and nomenclatural changes 40
    Species accounts 41
    References 42
ACKNOWLEDGEMENTS 44

SPECIES ACCOUNTS 45
    MARMOSETS, GOELDI'S MONKEY, TAMARINS AND LION TAMARINS · Callitrichidae 46
    SQUIRREL MONKEYS, GRACILE CAPUCHINS AND ROBUST CAPUCHINS · Cebidae 68
    NIGHT MONKEYS · Aotidae 81
    TITI MONKEYS, SAKIS, BEARDED SAKIS AND UACARIS · Pitheciidae 86
    HOWLER MONKEYS, SPIDER MONKEYS, WOOLLY MONKEYS AND MURIQUIS · Atelidae 111
    INTRODUCED OLD WORLD MONKEYS · Cercopithecidae 125

CHECKLIST OF THE NEOTROPICAL PRIMATES 126
INDEX 138

FOREWORD

I am honored to write the foreword to this new and comprehensive *Illustrated Checklist of Neotropical Primates*, compiled by some of the world's leading authorities on this amazing and diverse primate fauna. Indeed, the Neotropical region, one of the four major regions where wild primates are found, is the most globally robust and home to the highest number of species.

For the last 40+ years, I have worked closely with the authors to raise support for the comprehensive study of Central and South American and Mexican primates, document their distribution and status, and safeguard their future by initiating and developing a wide variety of conservation initiatives. Together, working under the auspices of the IUCN Species Survival Commission's Primate Specialist Group, we have been able to focus the world's attention on the threatened status of the region's hundreds of marmoset and monkey species, along with their threatened relatives in Africa, Asia and Madagascar.

At the turn of the current century, the Primate Specialist Group launched a list of the World's 25 Most Endangered Primates, which is reviewed and revised every two years, published as *Primates in Peril*, and serves as a guide for range country primatologists as to the degree of threat to wild primate populations and the opportunities for positive conservation action on their behalf. A significant amount of funding to support many of these initiatives, including those in Neotropical countries, has been provided by the Margot Marsh Biodiversity Foundation (established in 1996) and its corresponding Primate Action Fund (previously managed by Conservation International and now by Re:wild), and the Mohamed bin Zayed Species Conservation Fund (established in 2009). It has been my privilege to work with the authors in the creation, support and management of these funding mechanisms.

This new guide is the fifth in the series produced by Lynx, and the second sponsored by Re:wild, the first being *Mammals of Madagascar*. Besides serving as a guide to the what-and-where of the 217 Neotropical non-human primates, as well as highlighting conservation needs and strategies, it provides readily accessible information to any and all people interested in primate ecotourism, 'primate-watching' and 'primate life-listing', following the global model of birdwatching. Properly done, this kind of ecotourism can have huge positive impacts on the conservation of these animals and their tropical forest habitats, while also providing major benefits to local communities living close to nature.

Bill Konstant
Advisor, Margot Marsh Biodiversity Foundation
Advisor, Mohamed bin Zayed Species Conservation Fund
Senior Associate, Re:wild

INTRODUCTION

The Neotropical Region

The Neotropical Region (also sometimes referred to as the Neotropical Realm) is one of Earth's eight major zoogeographical regions. The name 'Neotropical' was first coined by Philip L. Sclater in 1858. His aim was to ascertain "the most natural primary divisions of the earth's surface, taking the amount of similarity or dissimilarity of organized life solely as our guide". Sclater's study was based on birds. The most influential work on defining zoogeographical regions, however, was that of the great explorer-naturalist Alfred Russel Wallace (Wallace, 1876). His classification of zoological regions followed that of Sclater, except for the name 'Indian', which he substituted with 'Oriental'. In his assessment of the Neotropical region, Wallace stated that it includes South America, the Antilles and tropical North America (southern Mexico), and "possesses more peculiar families of vertebrates and genera of birds and Mammalia than any other region". In simplest terms, it encompasses all of South America, including both the tropical and temperate portions, all of Central America, southern Mexico, the Caribbean islands and the southern tip of Florida.

The Neotropical primates

Neotropical primates occur in southern Mexico, Central America and South America, south to northern Argentina, Rio Grande do Sul (Brazil) and just into northern Uruguay, in all continental countries except for Chile. They are an extraordinary and diverse radiation of five primate families: marmosets, tamarins, lion tamarins and the unique Goeldi's monkey (Family Callitrichidae); squirrel monkeys and capuchin monkeys (Cebidae); night monkeys or owl monkeys (Aotidae); titi monkeys, sakis, bearded sakis and uacaris (Pitheciidae); and the howler monkeys, spider monkeys, woolly monkeys and muriquis (Atelidae). As of the publication of this checklist, they total 24 genera encompassing 192 species (217 species and subspecies) (Rylands & Mittermeier, 2024).

The Neotropical primates are monkeys of the Parvorder Platyrrhini, a name which refers to the shape of their noses (flat-nosed) with the nostrils well separated and pointing outward. It is a radiation quite distinct from the monkeys and apes of Africa and Asia, members of the Parvorder Catarrhini, which have noses with the nostrils in the shape of a hook close together and pointing downward. The Neotropical primates have also long been referred to as New World monkeys, as opposed to the Old World monkeys (Rosenberger, 2020), although there are arguments against using these names (Bezanson et al., 2024).

Evidence suggests that the platyrrhines diverged from the ancestors of the Asian and African monkeys, the catarrhines, about 45–36 million years ago (mya) in the middle to late Eocene Epoch following a trans-Atlantic dispersal from North Africa (Schrago & Russo, 2003; Schrago et al., 2014; Kay, 2015; Silvestro et al., 2019). A recent phylogenetic review indicated a slightly later date of 35.4 (33.9–37.1) mya (Beck et al., 2023). Fossil discoveries in the Andean foothills of Peru in western Amazonia have revealed two early ancestors from the late Eocene/early Oligocene, *Perupithecus* Bond et al., 2015, and *Ucayalipithecus* Seiffert et al., 2020, each with affinities to distinct lineages of early African primates (Oligopithecidae-like primates and Parapithecidae, respectively), suggesting two separate events for the colonization of South America by African anthropoids. In 2023, Marivaux et al. described a third fossil from the upper Rio Juruá in the Brazilian state of Acre, *Ashaninkacebus simpsoni*, which they identified as having strong affinities with a third distinct lineage of Asian-African stem anthropoids, the Eosimiidae. Silvestro et al. (2019) and Marivaux et al. (2023) indicated that these ancestral platyrrhines were small insectivore-frugivores weighing less than 400 g, similar to the marmosets and tamarins of today.

The platyrrhine fossil history is diverse, with about 39 genera (Fleagle et al., 1997; Tejedor & Novo, 2017; Rosenberger, 2020; Beck et al., 2023). Late Eocene and Oligocene fossils (36–23 mya) have been found in Bolivia, Peru and Brazil, and Miocene fossils (24–7 mya) in Colombia, Brazil, Peru, Chile and Argentina, south to the tip of Patagonia. The climate in Patagonia was then favourable, being forested prior to the Andean uplift that led to a cooler climate toward the end of the Miocene about 14 mya (Fleagle, 2000; Rosenberger, 2020). There are also 3–4 fossil genera known from the Greater Antilles, the islands of Cuba, Jamaica and Hispaniola but their identity and relation to the present day platyrrhines is still being discussed. For example, the fossil of the Jamaican genus *Xenothrix* is from the Holocene (about 11,000 years old) and while Kay (2015) argues that it is a so-called stem platyrrhine—a lineage older than that which gave rise to today's genera—Rosenberger (2020) sees it as a pitheciid, related to either the titi monkeys or night monkeys.

The Neotropical primates make up 30% of the 721 primates worldwide (Table 1). Brazil, with 23 genera and 151 species and subspecies (Table 2), leads the world in the number of primates since it includes a large portion of the Amazonian forests (nearly 6.9 million km^2), the largest river basin and expanse of tropical rain forest in the world and equivalent in size to the continental United States excluding Texas and California. The Congo basin comes second with 3.7 million km^2 (Goulding *et al.*, 2003; Fearnside, 2005). There are 150 species and subspecies of primates in Amazonia, which corresponds to 69% of the Neotropical primates and 21% of the world's living primates. In all, 124 of Brazil's primates are Amazonian but there are also 17 primates endemic to the Brazilian Atlantic Forest. Eighty-eight of Brazil's primates occur in no other country (endemics). Two other Amazonian countries rank high in the list of the world's primates: Peru is fifth with 58 (10 endemics) and Colombia sixth with 47 (14 endemics) (Table 2). Bolivia (23), Ecuador (23), and Venezuela (20) also have rich primate faunas (Table 3).

All the Neotropical primates are tropical or subtropical, forest dwellers and mainly arboreal. Only one species, the Mexican Spider Monkey, *Ateles geoffroyi vellerosus*, may have very marginally breached the Tropic of Cancer (23°30'N). This monkey has a range that extends up the east coast of Mexico into the dry forests of the state of Veracruz and west into southeast San Luis Potosí. Kellogg and Goldman (1944) suggested that it might even occur as far north as the south of the state of Tamaulipas in Mexico—a suspicion that has never been confirmed.

Seven possibly eight primates have ranges that extend south of the Tropic of Capricorn (23°30'S) in the southern Brazilian states of Paraná, Santa Catarina and Rio Grande do Sul, and into northern Argentina, southern Paraguay and the extreme north of Uruguay. The Black-faced Lion Tamarin, *Leontopithecus caissara*, and the southernmost populations of the Southern Muriqui, *Brachyteles arachnoides*, extend into northern Paraná. Azara's Night Monkey, *Aotus azarae azarae*, is found in the dry Chaco and humid Chaco of western Paraguay, east of the Río Paraguay, and in the provinces of Chaco and Formosa in northern Argentina. The Hooded Capuchin, *Sapajus cay*, extends into the far northwest of Argentina (Yungas) and the Atlantic Forest in Argentina, southeastern Paraguay, and marginally into the Paraguayan Humid Chaco, west of the Río Paraguay. The Southern Black-horned Capuchin, *Sapajus cucullatus*, occupies the Atlantic Forest in Rio Grande do Sul, Brazil, east of the Río Paraná, and Misiones Province, Argentina, south as far as 29°50'S. The southernmost populations of the Brown Howler, *Alouatta guariba*, are found in Misiones, Argentina, to 28°S and as far south as the Río Camaquã (31°10'S) in the state of Rio Grande

	Families	Taxa (spp. and ssp.)	Species	Genera
All Primates	16	721	539	83
Neotropics	5	217	192	24
Africa	4	198	107	26
Madagascar	5	110	106	15
Asia	5	196	134	19

Table 1. The number of primate species and subspecies worldwide and in the Neotropics, Africa, Madagascar and Asia.

	Country	Taxa	Species	Genera	Endemic taxa	Endemic species
1	Brazil	151	145	23	88	80
2	Madagascar	112	108	15	112	108
3	Indonesia	84	67	11	64	46
4	DR Congo	62	45	19	19	11
5	Peru	58	55	14	10	9
6	Colombia	47	43	16	14	12
7	Tanzania	46	28	13	15	8
8	Cameroon	37	32	19	0	0
9	Malaysia	34	26	9	6	1
10	China	32	28	9	10	6

Table 2. The 10 countries with the most primates. Two hundred and fifty primates occur only in these countries (country endemics), which represents 34% of all the currently recognized primate species and subspecies.

	Country	Taxa	Species	Genera	Endemic taxa	Endemic species	Threatened taxa	%
1	Brazil	151	145	23	88	80	53	35.3
2	Peru	58	54	14	10	9	17	29.8
3	Colombia	47	43	16	14	12	25	53.2
4	Ecuador	23	21	12	1	1	12	52.2
5	Bolivia	23	23	14	2	2	7	29.2
6	Venezuela	20	19	10	2	1	7	35
7	Panama	12	8	6	3	0	10	83
8	Guyana	9	9	8	-	-	-	11.1
9	French Guiana	8	8	8	-	-	1	12.5
10	Suriname	8	8	8	-	-	1	12.5
11	Costa Rica	6	4	4	1	-	6	100
12	Paraguay	5	5	5	-	-	1	20
13	Argentina	5	5	3	-	-	2	40
14	Nicaragua	5	3	3	1	1	4	100
15	Honduras	3	3	3	-	-	3	100
16	Guatemala	3	3	2	-	-	2	67
17	Mexico	3	3	2	1	-	3	100
18	Trinidad & Tobago	2	2	2	2	1	1	50
19	Belize	2	2	2	-	-	2	100
20	El Salvador	1	1	1	-	-	1	100
21	Uruguay	1	1	1	-	-	-	0

Table 3. The number of primate species and subspecies, genera, endemic taxa and endemic species in each of the 21 Neotropical countries where they occur. The columns "Threatened taxa" and "%" give the number and the percentage of the species and subspecies that are threatened (Critically Endangered, Endangered and Vulnerable) in each country.

Sul, Brazil, where the Atlantic Forest gives way to the Pampas Grasslands. The range of the Black and Gold Howler Monkey, *Alouatta caraya*, extends through the Humid and Dry Chaco and Atlantic Forest in Paraguay, western parts of the states of Paraná, Santa Catarina and Rio Grande do Sul, the provinces of Misiones, Corrientes, Chaco, eastern Formosa and northern Santa Fe in Argentina, and the extreme north of Uruguay (Jardim *et al.*, 2020). It is possible that the Pale Titi, *Plecturocebus pallescens*, extends just south of the Tropic of Cancer in central Paraguay. The southern limits of its distribution in the Cerrados del Chaco and humid areas in the dry Chaco of Paraguay are not well known, and the southernmost known record to date is only a few minutes north of the Tropic of Capricorn (Smith *et al.*, 2021).

Diversity

Today, 59 species (64 species and subspecies) of tamarins and marmosets, are recognized in 10 genera of the Family **Callitrichidae** (Table 4). The tamarins, until recently all placed in the genus *Saguinus*, are now divided into four genera: the white-mouthed group, *Leontocebus*, in central and western Amazonia; the moustached tamarins, *Tamarinus*, also in central and western Amazonia, the Golden-handed Tamarin and Brazilian bare-face tamarins, *Saguinus*, on the Guianan and Brazilian shields in eastern, northern and north-eastern Amazonia; and the Panamanian and northern Colombian tamarins, *Oedipomidas* (Rylands *et al.*, 2016, Brcko *et al.*, 2022). There are four genera of marmosets: the pygmy marmosets, *Cebuella*, from western Amazonia; the dwarf marmoset, *Callibella*, restricted to a small region in central Amazonia; the largely Amazonian marmosets of the genus *Mico*; and the eastern Brazilian marmosets, *Callithrix*. The monotypic genus *Callimico*, Goeldi's Monkey, related to the marmosets, is found in western Amazonia. Lastly, there are four species of lion tamarins, *Leontopithecus*, in the Atlantic Forest of eastern Brazil. They are now known to represent an early lineage that separated from the ancestral marmosets, so should really be called 'lion marmosets' rather than 'tamarins' (Coimbra-Filho & Mittermeier, 1972); nevertheless, we do not recommend changing this now, given that the name 'lion tamarin' has been in use for more than 50 years.

	Species	Taxa	Endangered (CR+EN)	%	Threatened (CR+EN+VU)	%
All Neotropical Primates	192	217	52	24.4	94	44.1
Callitrichidae	59	64	9	14.1	19	29.7
Cebidae	33	37	14	37.8	20	54
Aotidae	11	13	1	7.7	6	46.1
Pitheciidae	63	63	9	14.3	22	34.9
Atelidae	26	40	19	47.5	28	70

Table 4. The number of endangered (CR and EN) and threatened (CR, EN and VU) primates in each of the five families of Neotropical primates.

The chief features of the callitrichids that separate them from other platyrrhines are their small size (at 110–700 g, the smallest of the Neotropical monkeys), their claw-shaped nails on all digits except for the big toe, two rather than three molar teeth on each side of each jaw, and their propensity to give birth to dizygotic twins (Ford et al., 2009). Goeldi's Monkey, *Callimico goeldii*, an early offshoot of the marmoset lineage, is the only exception as it has three molars (the third is small) and gives birth to singletons. Weighing 110–125 g, the pygmy marmosets, are the smallest of all monkeys. All the callitrichids are fruit-eaters and predators of small animals, including arthropods, snails, frogs, lizards, birds' eggs and nestlings. All also eat plant exudates, i.e. nectar and gums and, rarely, terpenoid resins and latex. The marmosets are specialist gum-feeders, fermenting it and digesting it in an enlarged caecum, and have morphological adaptations (notably large rodent-like incisors and canines) for gouging tree trunks, branches and vines to stimulate the excretion of gum. The tamarins of Amazonia, northern Colombia and Panama, *Oedipomidas*, *Leontocebus*, *Tamarinus* and *Saguinus*, the lion tamarins, *Leontopithecus*, of the Brazilian Atlantic Forest and Goeldi's Monkey eat gum only opportunistically. Fungi are eaten by a few species but are a significant component of the diet of Goeldi's Monkey and some populations of the two Atlantic Forest montane marmosets, *Callithrix aurita* and *C. flaviceps*. Marmosets and Goeldi's Monkey generally breed twice a year, while the tamarins and lion tamarins usually do so just once. Callitrichids are unusual among primates—and even among mammals—in several aspects of their mating and reproductive behaviour. They have a variable mating system, ranging from monogamy to polygyny and polyandry, and even polygynandry (two or more males mate with two or more females), depending on the group size (two to about 20) and composition. Generally, only one female in each social group breeds; she suppresses breeding by other females behaviourally and/or physiologically through pheromones from her scent glands. The relatively large twin infants are carried by other group members (especially adult males) shortly after birth and are fed by them whilst they are weaning but still lacking the competence to capture prey. They become independent at about five months. As such, callitrichids are cooperative breeders—the entire group carries, feeds and protects the breeding female's offspring. The callitrichids are genetically close to the Cebidae (squirrel monkeys and capuchins) and are considered by some to be a subfamily of cebids, the Callitrichinae, rather than a family of their own.

The Family **Cebidae** currently has three genera, 33 species and 37 species and subspecies. They range from Central America (Costa Rica, Honduras, Nicaragua and Panama) through northern South America, south to northern Argentina, Bolivia, Paraguay and southern Brazil. At 750–1100 g, the squirrel monkeys, *Saimiri*, are a little larger than the callitrichids. The capuchin monkeys are considerably larger, averaging 3 kg, and large robust males can reach 4.8 kg. Some of the squirrel monkey species are seasonally sexually dimorphic—the males fatten up during the mating season to become about 20% heavier than the females. Male capuchins are, in general, about 24% larger than females. All three genera (*Saimiri*, *Cebus* and *Sapajus*) have large brains relative to their body size. Squirrel monkeys are frugivores but also highly insectivorous. Large groups (sometimes of more than a hundred) travel 2–5 km a day, foraging for insects on the leaves and branches of the forest understorey while searching for fruiting trees with large quantities of small fruits such as mast-fruiting fig trees.

Formerly placed in a single genus *Cebus*, two genera of capuchin monkeys are now recognized: *Cebus*, the gracile capuchins, and *Sapajus*, the robust capuchins (Fragaszy et al., 2004; Lynch et al., 2012). They are also insectivore-frugivores but, being much larger than most insectivorous primates, they have a broader diet including larger fruits that are often tougher, and even hard-shelled. This is particularly the case of the robust capuchins, *Sapajus*, that include especially palm fruits in their diets. The gracile capuchins, *Cebus*, also eat palm fruits but generally softer fruits. The capuchins are manipulative, destructive, extractive foragers that pull, rip, bite, puncture and smash potential food items. They have a semi-prehensile tail acting as a fifth limb that they wrap around a branch for extra support; with the hindlimbs it acts as a tripod, thereby freeing up their hands for foraging and manipulating fruits (Garber & Rehg, 1999). Squirrel monkeys cannot move their digits independently and so can only grab and grasp their prey and small fruits. Groups of up to 30 capuchins travel and forage in higher strata of the forest than the squirrel monkeys. The disturbance caused by their foraging can attract followers such as Nunbirds (puff-birds of the genus *Monasa*) and Double-toothed Kites (*Harpagus bidentatus*), which take advantage for their own foraging. In Amazonia, the robust capuchins and to a lesser extent the gracile species can often be found travelling with squirrel monkeys, the latter benefiting from the disturbance and pickings dropped and disturbed by the capuchin monkeys from the canopy above. The capuchins benefit from the squirrel monkeys' vigilance for predators in the understorey, while the squirrel monkeys benefit from the capuchins' alertness for aerial predators such as hawks and eagles. Capuchins have pseudo-opposable thumbs that make them very dexterous. Robust capuchins use tools in a variety of ways, often similar to the tool use shown by chimpanzees. Some populations, for example, use stones to break open hard palm fruits, to dig out and break up tubers, to excavate for insects, and to open hollow branches to obtain larvae and ants. Twigs or branches, used to probe holes in trees and rock crevices for insects, honey and water, are often modified for this purpose. The capuchins in the drier regions of north-eastern Brazil, the Bearded Capuchin, *Sapajus libidinosus*, and the Yellow-breasted Capuchin, *S. xanthosternos*, use large heavy stones to smash open hard palm fruits. Panamanian White-faced Capuchins, *Cebus imitator*, on Coiba Island, Panama, use stones to break open seeds and crab and snail shells (Barrett *et al.*, 2018). The capuchins' precision grip between their thumbs and second digits (and between their second and third digits) is not found in any of the other platyrrhines. Capuchins are also unique for their delayed maturation and long lifespan, their coalitionary behaviour and, in gracile capuchins, their social traditions including game-playing that is believed to test affiliative bonds.

We place the night monkeys (also known as owl monkeys) in their own family, **Aotidae**, for expediency due to the anomaly that their morphology and behaviour places them close to the titi monkeys in the Pitheciidae (Subfamily Homunculinae) (Rosenberger & Tejedor, 2023), whereas phylogenetic analyses place them in the Cebidae, along with the capuchins and squirrel monkeys, as the Subfamily Aotinae (Schneider & Rosenberger, 1996; Rosenberger, 2020). They retain a generalized morphology and behaviour similar to the titi monkeys in the Pitheciidae, but phylogenomic analyses place them as the sister group to the Callitrichinae (Kuderna *et al.*, 2023). There is just one widespread genus, *Aotus*, with 11 species and 13 species and subspecies (Fernandez-Duque, 2023), extending from Panama and Colombia through Amazonia, south to Bolivia, Paraguay and northern Argentina but absent from the Atlantic Forest and the Guianas.

Night monkeys are socially pair-bonded, and males and females are similar in weight (0.7–1.2 kg) and appearance. They are the only nocturnal monkeys. They have large, spherical eyeballs, with lower densities of cone cells (those specialized for daytime vision), and very high densities of rod cells in the central retina (specialized for vision in low light intensity). One of the larger night monkeys, *A. azarae azarae* of the Argentinean and Paraguayan Chaco, has been found to be cathemeral—active for periods at night and in the day. The small family groups, of up to six individuals—a breeding pair and their offspring—rest together, usually in holes in trees or dense vine tangles, during the day. They are largely frugivorous, but also eat leaves, arthropods (e.g., moths, beetles, spiders, millipedes and cockroaches), flowers and fungi. They have single offspring, and the male is active in caring for and carrying the infant.

There are six genera and 63 species of titi monkeys, sakis, bearded sakis and uacaris in the Family **Pitheciidae** (Veiga *et al.*, 2013; Norconk, 2020). They are very largely Amazonian in their distribution, with outliers extending to the Orinoco, and a genus of titi monkeys, *Callicebus*, occurring in the Atlantic Forest of eastern Brazil, as well as in forest patches in the Caatinga of North-east Brazil (*Callicebus barbarabrownae*) and the Cerrado of eastern Brazil (*C. nigrifrons*). The Pale Titi (*Plecturocebus pallescens*) occurs in the Chaco of Bolivia and Paraguay and the margins of the Brazilian Pantanal. They are all, to varying degrees, seed predators. They can extract soft seeds from fruits with hard pericarps (Kay *et al.*, 2013). They do this with a specialized dentition (notably, large, splayed canines and high-crowned procumbent incisors), well-developed temporal (biting) and masseter (chewing) muscles, and deep jaws (Kay *et al.*, 2013). These adaptations are increasingly pronounced from the titi monkeys, *Callicebus*, *Plecturocebus* and *Cheracebus*, to the sakis, *Pithecia*, and are most extreme in the bearded sakis, *Chiropotes*, and uacaris, *Cacajao*.

The smallest members of the family, the titi monkeys, range in size from 700 g to 1.5 kg. Formerly, they were considered to belong to a single genus, *Callicebus*, but based on molecular, morphological and biogeographic evidence, Byrne *et al.* (2016) proposed a new genus-level taxonomy: *Cheracebus* for the widow monkeys or collared titis in the Orinoco, Negro and upper Amazon basins (*torquatus* group); *Plecturocebus* for the titis of the Amazon basin and Chaco region, and the edges of the Pantanal, in what would be Cerrado; and *Callicebus* for the five titis in the Atlantic Forest, Caatinga and Cerrado of eastern and northeastern Brazil (*personatus* group). They live in family groups consisting of an adult pair and their offspring. They can generally be seen in the understorey and the lower strata of the forest up to 10 m, although the Atlantic Forest titis, *Callicebus*, can spend a considerable amount of time in the lower canopy of the taller forests and so are less easy to find. Titis eat fruits and seeds from tough sclerocarpic fruits, besides leaves, flowers and insects (Bicca-Marques & Heymann, 2013). In western Amazonia, titis of the genera *Plecturocebus* and *Cheracebus* can be sympatric, and it is argued that the latter is more of a seed predator than the former. Although seeds can make up more than a third of the diet of collared titis, the few studies of members of the genus *Plecturocebus* have found less seed predation and a more folivorous diet. These monkeys have small home ranges, as little as 1 ha but in some cases up to 25 ha. They are territorial and have distinctive calling sessions including duets of the mated pair that usually take place in the early morning.

The sakis, genus *Pithecia*, are entirely Amazonian and range in weight from 1.4 to 3.2 kg. Some species such as *P. pithecia* forage and travel largely in the lower and middle forest canopy, which correlates with their predominantly vertical clinging and leaping locomotion, whereas other larger species such as *P. albicans* and *P. irrorata* use more quadrupedal walking and jumping in the middle and even upper canopy. Most of the species are sexually dichromatic—females and males having subtle or distinct differences in their pelage coloration. Group sizes are variable, but in general between two and six individuals, exceptionally up to eight. Seeds make up 26–64% of the annual diet, along with fleshy fruit and entire fruits, young leaves and flowers (Norconk & Setz, 2013).

Bearded sakis, *Chiropotes*, vary in size from 2.0 to 4.0 kg, with males being a little larger than females. Males are remarkable for the pom-pom swellings of the enlarged temporal muscles on their heads, their bulbous beards—longer on adult males—and their bushy tails, which they tend to wag when disturbed or excited. They are entirely Amazonian, occurring in the *terra firme* forests of the lower Amazon basin, east of the rios Madeira, Jiparaná and Negro, extending north to the middle and lower Orinoco (Veiga & Ferrari, 2013). They live in large multimale-multifemale groups ranging from 15 to 65 members and travel widely (home ranges can exceed 500 ha) in search of tall trees with large fruit crops, especially those in the families Lecythidaceae (of the Brazil nut) and Sapotaceae (Ayres & Prance, 2013).

Weighing up to 4 kg, the uacaris are found west of the rios Negro and Purus in the upper Amazon in Brazil, Colombia and Peru. There are two types: the bald uacaris, mostly inhabitants of the white-water seasonally flooded forests of western Amazonia, south of the rios Japurá and Amazonas/Solimões; and the black-headed uacaris of the black-water seasonally inundated forests of the Guiana Shield in the basins of the Rio Negro and upper Orinoco. The uacaris have long shaggy coats and, unlike all other Neotropical monkeys, short tails. They can form larger groups than even the bearded sakis. Group sizes range from 20 to 200 members. Depending on the abundance and dispersion of their fruit trees, uacaris tend to what is termed 'fission-fusion', with groups splitting temporarily to forage separately. Depending on the season and availability of food they may migrate out of the flooded forest to occupy *terra firme* forest. Some populations live exclusively in *terra firme* forest. Their diets are similar to those of the bearded saki—seeds, mostly unripe, are predominant, with ripe fruit, new leaves and flowers being supplementary. Like the bearded sakis, the uacaris are seed predators more than seed dispersers. Both genera have enlarged canines to puncture and rip open hard fruits, and procumbent incisors that act as pincers to extract the seeds (even small seeds 0.25 cm long) that are triturated by their low crowned molars (Ayres & Prance, 2013; Barnett *et al.*, 2013). One of the mysteries of Neotropical primate evolution is why the bearded sakis and the uacaris, so similar in dietary preferences and behaviour, have such different tails.

The largest of the Neotropical monkeys are those in the Family **Atelidae**, with four genera, 26 species and 40 species and subspecies. The family includes the howler monkeys, *Alouatta*, in the Subfamily Alouattinae, the most wide-ranging of the Neotropical monkeys, occupying forests from southern Mexico, south to northern Argentina and southern Brazil. In the Subfamily Atelinae are the spider monkeys, *Ateles*, of southern Mexico and Central America, western Colombia and Ecuador, and throughout Amazonia; the woolly monkeys, *Lagothrix*, in Colombia and western Amazonia, west of the Rio Madeira; and the muriquis, *Brachyteles*, endemic to the Atlantic Forest of south-eastern and southern Brazil, extending just a little into the north-east in the state of Bahia, and south in the state of Paraná. The Central American Black Howler, *Alouatta pigra*, the Red-faced Black Spider monkey, *Ateles paniscus*, of the Guiana Shield and the woolly monkeys weigh around 7–10 kg. Male howlers and woolly monkeys are larger than females. The two species of muriqui are believed to be the largest of all, with females weighing up to 12 kg and males nearly 14 kg (Ruschi, 1964).

The atelids travel through the forest canopy using quadrupedal walking, climbing and clambering, and the atelins (spider monkeys, woolly monkeys and muriquis) also use suspensory locomotion. They have a prehensile tail with a bald patch towards the tip which has friction ridges. It is called the volar pad and gives them the grip needed to hang by their tails, which, as such, serves as a fifth limb (Organ *et al.*, 2011). While all use their tails when feeding, hanging from a branch to free up their hands to grasp leaves and fruits, the atelins are also brachiators—swinging by their arms but with the assistance of their tail, which differentiates them from the tailless Asian gibbons. The atelins have long limbs, increased mobility in the wrist, and long fingers that have curved phalanges for grasping branches and vines. Woolly monkeys differ from the spider monkeys and muriquis, however, in that they rely much more on quadrupedal locomotion than suspensory locomotion for travel and feeding (Defler, 1999).

The howler monkeys are the most folivorous of the Neotropical primates (Kowalewski et al., 2015). The males are larger than the females and in some species the sexes have distinct pelage coloration. The groups of 5–15 individuals (sometimes more) are generally made up of a dominant male, younger and subordinate males, 2–5 females, and a number of juveniles and infants. Their folivory is associated with relatively small home ranges of 5–20 ha. Depending on the season and the floristic composition of the forest, 50–80% of their diet is made up of mostly new leaves that have more protein and less fibre. They also eat fruits, seeds and flowers. Fruit increases in importance in their diets in areas with higher rainfall where groups are larger and have more extensive home ranges (Dias & Rangel-Negrín, 2013). Their very loud howling—which sounds more like roaring—is amplified by a soundbox formed from the hyoid bone in the throat. The soundbox of the male is larger than that of the female. The hyoid apparatus of male red howlers is the largest and they are the loudest. Bouts are generally heard around dawn but can occur at any time of day or even the night. The dominant male initiates a session and is joined by the females and other group members. Howling is believed to serve as a territorial marker (although home ranges can overlap extensively) and as a display of strength by the dominant male as he defends his group from the constant threat of takeover by other males. Dominance struggles can result in vicious fights. On reaching maturity, both males and females disperse from their natal groups.

Unlike the howler monkeys, the spider monkeys, *Ateles*, are highly frugivorous and fruits make up 64–100% of their diet in any given month (Di Fiore et al., 2008). They require large home ranges of hundreds of hectares (in some cases nearly 1000 ha) to guarantee sufficient trees in fruit throughout the year. They are important seed dispersers and for many trees and vines with large seeds they are the *only* dispersers. Spider monkey groups can have 20 or more individuals but are rarely all seen together. Males sometimes form large subgroups. Small subgroups of variable size and composition travel separately to exploit widely distributed fruiting trees in patches that would not accommodate a large group. Larger subgroups are more common during periods of greater fruit abundance. Again, this social dynamic is called 'fission-fusion'. Males are philopatric, that is, they tend to stay in their natal groups and collaborate in defending their groups of females and juveniles from intruding males. Females disperse to other groups when they are sexually mature.

The stockier woolly monkeys, *Lagothrix*, are also highly frugivorous but include more leaves and insects in their diet than is typical in spider monkeys (Defler & Stevenson, 2014). For this reason, their large groups, up to 50 members with multiple adult males and females, are more cohesive, only occasionally splitting temporarily into two groups to search for fruit when it is scarce. Their home ranges are also large, extending in some cases to as many as 1000 ha, and they can travel 3 km or more in a single day.

The two species of muriquis, *Brachyteles*, in the montane Atlantic Forest are also frugivores but, like the woolly monkeys, also include more leaves, especially tender young leaves and buds, in their diet. They live in large groups of as many as 40–50 individuals, with philopatric males and females dispersing to other groups when they reach puberty. In large expanses of forest, home ranges can be as large as 1500 ha. In richer forests and small forest patches, groups can exceed 100 members and accordingly become less cohesive, forming subgroups that when travelling spread out over 100–300 m. Muriquis are well known for their egalitarian social organization and for the large group hugs that occur among related males (Strier, 1999).

New species

Primate surveys and taxonomic revisions have resulted in the description of 104 new species and subspecies of primates since 2000: 52 from Madagascar, nine from Africa, 19 from Asia, and 24 from the Neotropics. All but one of the Neotropical discoveries were Amazonian: four marmosets of the genus *Mico*, two tamarins (*Leontocebus*), nine titi monkeys (eight of the genus *Plecturocebus* and one of the genus *Cheracebus*), five sakis (*Pithecia*) and three uacaris (*Cacajao*). A distinctive night monkey (*Aotus jorgehernandezi*) was discovered in northern Colombia but very little is known about it, since its description is based on just a single individual.

Three mystery monkeys

There are three mysterious primates, two of which are not listed here because of uncertainties regarding their existence. The first is an Amazonian tamarin that was kept in the Bronx Zoo, New York, from 1951 to 1954. It was an adult male and was described by Hershkovitz in 1966 from the skin and skeleton preserved in the American Museum of Natural History. Hershkovitz considered it to be a subspecies of Spix's Saddle-back Tamarin and named it Crandall's Saddle-back Tamarin, *Saguinus fuscicollis crandalli*. Lee Crandall was the curator of the Bronx Zoo at the time. It was certainly distinctive, with a broad white blaze across its forehead, a buff-to white throat, neck and chest, orange on the rump and thighs, drab (dull, pale brown) on the crown, mantle and arms, and a pale brown tail. Its provenance is unknown and no tamarin like it has been seen since. Hershkovitz (1977) indicated that it might have come from the upper reaches of the Purus and Juruá basins. The possibility remains, however, that it was a hybrid. Hershkovitz (1977) noted that it was intermediate in coloration between the darker *Saguinus fuscicollis cruzlimai* (here called *Leontocebus cruzlimai*) from the upper Purus and the paler *S. f. acrensis* described by Carvalho in 1957 from the upper Rio Juruá. The Acre Saddle-back Tamarin was subsequently found to be a hybrid *Leontocebus fuscicollis fuscicollis* × *Leontocebus fuscicollis melanoleucus* (see Peres et al., 1996).

The second mystery is a recently described titi monkey, *Plecturocebus parecis* Gusmão et al., 2019, from the central southern part of Brazilian Amazonia in the state of Rondônia, named after the Chapada dos Parecis where it was found. It is grey, with a reddish-brown back, and an off-white beard and sideburns. It is closely related to another titi, *P. cinerascens*, described by Spix in 1823, and aptly called the Ashy Titi, that also occurs in the basin of the upper Rio Madeira but further north. Byrne *et al.* (2021) reviewed the evidence for *P. parecis* being a distinct species and concluded that there was a strong argument for it being considered just a cline of *P. cinerascens*, with gradual variation of pelage coloration (the whitish beard and the reddish-brown back) from the northern to the southern populations. Gusmão *et al.* (2019) and Byrne *et al.* (2021) agreed that further study is required.

The third is a spider monkey. The Grizzled or Hooded Spider Monkey, *Ateles geoffroyi grisescens* (Sclater in Gray, 1866) (see page 119) was described in a manuscript that catalogued the vertebrates in the London Zoological Gardens in 1865. Gray (1866) quoted Philip Sclater's diagnosis of the 'Grizzled Spider Monkey': "Fur moderately long, black, with many silvery-white hairs interspersed; tail black, underside greyish; hair of the forehead moderately long; face – ?; thumb none". He attributed the authorship to Sclater. The type is preserved in the British Museum of Natural History, London (Gray, 1866; Napier, 1976). The type locality, however, is unknown. Kellogg and Goldman (1944) reviewed the subspecies of the Central American Geoffroy's Spider Monkeys and suggested that, even though they had not seen the type specimen of *A. g. grisescens*, it might hail from the Río Tuyra (Tuira) basin, Panama. Hernández-Camacho and Cooper (1976) indicated that it also occurred in Colombia, extending south-eastward through the Serranía del Sapo in far south-east Panama and into the Cordillera de Baudó in north-west Colombia. These authors provided a description that differs from that of Sclater (in Gray, 1866): "characterized by a brownish or rusty-colored back (with black hair tips) and by completely black head, limbs, and tail" and suggested it might be a transitional form between the brightly coloured *A. geoffroyi* group of middle America and the largely jet-black *Ateles rufiventris* of southeast Panama and adjacent Colombia. In her 1976 catalogue of the Neotropical primates in the British Museum of Natural History, Napier placed the form *Ateles rufiventris* Sclater, 1872, as a synonym of *grisescens* (the former was described after the latter). Méndez-Carvajal and Cortés-Ortiz (2020) affirmed that *A. g. grisescens* had never yet been seen in the wild. Méndez-Carvajal (2021) reported, however, that a group of black spider monkeys with a fringe of whitish hairs on the chin and cheeks had been found on the Pacific side of eastern Panama and believed it to be the spider monkey that Sclater had described. Its whitish hairs around the chin and mouth are, however, a diagnostic feature of the Colombian Black Spider Monkey, *Ateles rufiventris*, the range of which extends into southern Panama (Kellogg & Goldman, 1944). This is not the whole story of the confusion regarding its identity, and the resolution of the validity or otherwise of this named taxon will require surveys to prove its existence in its presumed range and genetic analyses to see if it is a distinct and definable subspecies of Geoffroy's Spider Monkey or, as suggested by Napier (1976), a senior synonym of what we currently call *A. rufiventris*.

Quaternary fossil primates

In the Holocene and Pleistocene, there were native platyrrhine species on some Caribbean islands, as evidenced by fossils from Jamaica (*Xenothrix mcgregori*), Hispaniola (*Antillothrix bernensis* – Dominican Republic; *Insulacebus toussaintiana* – Haiti), and Cuba (*Paralouatta varonai* and *P. marianae*) (Cooke *et al.*, 2011, 2016; Rosenberger, 2020). All were quite large primates that ranged in weight from an estimated 4–5 kg (*Insulacebus*) to 7–9 kg (*Paralouatta*). Establishing the affinity of these fossils to the present day continental platyrrhines is no easy task. It is largely based on the morphology of some few teeth, mandibles, skulls and skull fragments, and is still subject to much measurement, comparison and conjecture. The genus *Paralouatta* was thought to be affiliated with the present-day howler monkeys, although some now argue its affinity, despite its name, to the titi monkeys. *Antillothrix bernensis* was at one time placed in the genus *Saimiri* (squirrel monkeys) but today it is thought that it could be related to the capuchin monkeys (Subfamily Cebinae). It has been suggested that *Insulacebus* might be related to *Xenothrix* and that both may have a phylogenetic affiliation with the genus *Aotus* (the night monkeys) or even the titi monkeys (Subfamily Callicebinae). *Xenothrix*, the first fossil primate found in the Caribbean, was so different from other platyrrhines that Hershkovitz (1970) controversially placed it in its own family, Xenothricidae. Post-cranial bones indicate that it was a "non-leaping, non-swinging, slow-moving arborealist", somewhat convergent with lorises and sloths (McPhee & Horovitz, 2002). All these Caribbean fossil primates date from the Holocene (*Xenothrix*) and Pleistocene except *Paralouatta marianae*, which is suspected to have a Miocene origin even though Beck *et al.* (2023) indicated a date of 1.29 mya in the second half of the Quaternary. Radiocarbon dating indicated that one of the *Xenothrix* fossils was alive only 900 years ago (Cooke et al., 2017) and could have been extirpated by hunting or the introduction of the African yellow fever virus.

Three large Late Pleistocene sub-fossils of the Family Atelidae (howler monkeys, spider monkeys, woolly monkeys and muriquis) were discovered in caves at two sites in Brazil. *Protopithecus brasiliensis* was a large atelid, somewhat larger than but similar to the muriqui. A femur and a humerus were discovered by Peter Wilhelm Lund in 1936 in the Lapa de Periperi, a cave near Lagoa Santa in the state of Minas Gerais, and named by him in 1938.

Two almost complete skeletons were found in another cave in 1993, the Toca de Boa Vista in the state of Bahia in North-east Brazil, about 1200 km north of Lagoa Santa. One was initially thought to belong to the genus *Protopithecus* but a more discerning morphometric analysis showed it to be rather different. It was described by Cartelle and Hartwig (1996) and given the name *Caipora bambuiorum*. The mandible and dentition clearly align it with the spider monkeys (*Ateles*) and the limb proportions and robust skeleton indicate adaptations for suspensory postures and brachiation. Its weight was estimated at 20 kg. The other skeleton, also an atelid, was larger still, with an estimated body weight of 25–28 kg, more than twice the weight of any extant platyrrhine. It was described by Halenar and Rosenberger (2013) and given the name *Cartelles coimbrafilhoi*. The cranium and skeleton are howler-like but the dentition is clearly more adapted to a frugivorous and seed-eating diet. The skeleton has long limbs, a long tail, large hands and feet, and mobile joints, which Rosenberger (2020) considered to be "basically designed for quadrupedal climbing and clambering, suspensory and, possibly, tail-assisted locomotion".

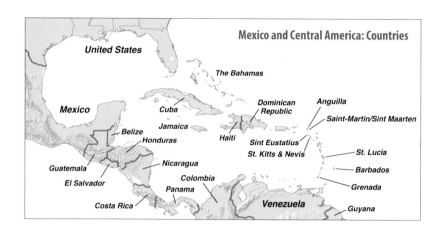

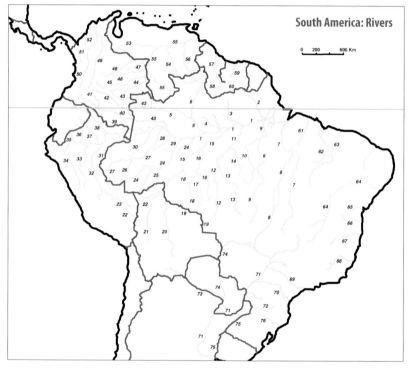

1. Rio Amazonas; **2.** Rio Jari; **3.** Rio Trombetas; **4.** Rio Uatumã; **5.** Rio Negro; **6.** Rio Branco; **7.** Rio Tocantins; **8.** Rio Araguaia; **9.** Rio Xingu; **10.** Rio Iriri; **11.** Rio Tapajós; **12.** Rio Juruena; **13.** Rio Teles Pires; **14.** Rio Jamanxim; **15.** Rio Madeira; **16.** Rio Aripuanã; **17.** Rio Roosevelt; **18.** Rio Jiparaná **19.** Rio Guaporé; **20.** Rio Mamoré; **21.** Río Beni; **22.** Río Madre de Dios; **23.** Río Manu; **24.** Rio Purus; **25.** Rio Tapauá; **26.** Rio Envira; **27.** Rio Juruá; **28.** Rio Tefé; **29.** Rio Coarí; **30.** Rio Jutaí; **31.** Río Javarí; **32.** Río Ucayali; **33.** Río Huallaga; **34.** Río Marañón; **35.** Río Santiago; **36.** Río Pastaza; **37.** Río Tigre; **38.** Río Napo; **39.** Rio Içá/Putumayo; **40.** Rio Japurá/Caquetá; **41.** Río Caguán; **42.** Río Apaporis; **43** Río Vaupés; **44.** Río Inirida; **45.** Río Guaviare; **46.** Río Vichada; **47.** Río Tomo; **48.** Río Meta; **49.** Río Magdalena; **50.** Río Cauca; **51.** Río Atrato; **52.** Río Sinu; **53.** Río Apure; **54.** Río Caura; **55.** Río Orinoco; **56.** Río Caroni; **57.** Essequibo River; **58.** Rupununi River; **59.** Coppename River; **60.** Suriname River; **61.** Rio Gurupí; **62.** Rio Itapecuru; **63.** Rio Parnaíba; **64.** Rio São Francisco; **65.** Rio Paraguaçú; **66.** Rio de Contas; **67.** Rio Jequitinonha; **68.** Rio Doce; **69.** Rio Tieté; **70.** Rio Paranapanema; **71.** Rio Paraná; **72.** Río Iguácu; **73.** Río Pilcomayo; **74.** Río Paraguai; **75.** Río Uruguai.

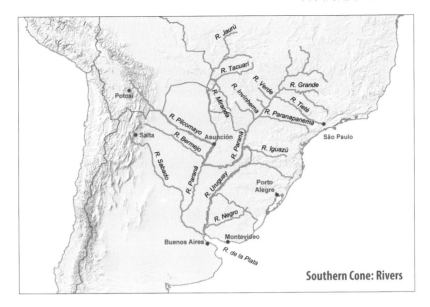

Southern Cone: Rivers

Their habitats

The distributions of the Neotropical primates are quite regionalized according to the topography and climate and the vegetation types and their floristic composition that provide the habitats for these monkeys. They can be broadly defined as: (1) southern Mexico and Central America and west of the Andean Cordillera along the Pacific Coast, through Ecuador to northern Tumbes in Peru; (2) central and northern Colombia, northern Venezuela and Trinidad; (3) the foothills and montane and cloud forests of the Andes in southern Colombia, Ecuador, Peru and Bolivia; (4) Amazonia; (5) the Brazilian Cerrado and Caatinga, two distinct biomes but primate-species-poor and so here combined; (6) the Atlantic Forest of Brazil, northern Argentina and south-eastern Paraguay; and (7) the Southern Cone countries including northern Argentina, southern Bolivia, Paraguay and northern Uruguay.

1. Southern Mexico and Central America and west of the Andean Cordillera Occidental in Colombia, Ecuador and north-western Peru (Tumbes)

There are three major forest types. **Tropical Dry Broadleaf Forest** predominates in the southern part of eastern Mexico, in north and central Veracruz and the north of the Yucatán Peninsula (Yucatán Province), and extends along the Pacific coast of Central America west of the Sierra Madre as far south as the Nicoya Peninsula in Costa Rica. The forest changes to **Tropical Moist Broadleaf** in Mexico, in southern Veracruz, northern Oaxaca and Chiapas, and Tabasco, into Campeche and Quintana Roo provinces, Belize and Guatemala on the Yucatán Peninsula, east (north and east of the Sierra Madre) through eastern Guatemala, northern Honduras and northern and eastern Nicaragua, through Costa Rica and Panama, and extending south and west of the Andean Cordillera along the Pacific coast in the Chocó biogeographic region of Colombia and into Ecuador in the provinces of Esmeraldas and Manabí. Dry forest predominates south into Tumbes in northern Peru. Depending on the elevation, primates can also occupy **Montane** and **Cloud Forest** (above 1500 m) in the Sierra Madre and Andes up to 2500 m.

> *Notable species:* Geoffroy's Tamarin (*Oedipomidas geoffroyi*), the Central American Squirrel Monkey (*Saimiri oerstedii*), white-faced capuchins (*Cebus imitator* and *C. capucinus*), the Ecuadorian White-fronted Capuchin (*Cebus aequatorialis*), the Panamanian Night Monkey (*Aotus zonalis*), the Mantled Howler (*Alouatta palliata*), the Central American Black Howler (*Alouatta pigra*), Geoffroy's Spider Monkey (*Ateles geoffroyi*) and the Ecuadorian and Colombian spider monkeys (*Ateles fusciceps* and *A. rufiventris*).

2. Central and northern Colombia, northern Venezuela and Trinidad

Colombia is divided into four major topographical regions. Besides the Pacific coast (a largely lowland strip but with isolated peaks and intermittent extensions of the Western Cordillera), the Andean region comprises the three parallel Andean cordilleras: Occidental (up to 3000 m), Central (the highest up to 5500 m) and Oriental (the widest and up to 5300 m). They are separated by the **Submontane** and

Montane (over 700 m elevation) **Moist Forests**, **Cloud Forests** (above 1500 m) and the **Dry Forests and Humid Forests** of the valleys of the ríos Cauca and Magdalena. The Caribbean Coastal plain (*Llanura Costera del Caribe*) in north Colombia, east from the Río Sinu (just north of the Panamanian border) is hilly in the south with **Tropical Moist Broadleaf Forest** changing to what are now just small remnants of the once widespread **Dry Broadleaf Forest** towards the north. Isolated mountains on the Caribbean coast have patches of moist tropical forest, although the Guajira Peninsula is arid. East of the Andes in Colombia are the *Llanos Orientales*, with **Gallery Forest** along the left bank (western) tributaries of the Río Orinoco, including the ríos Arauca, Meta, Tomo and Vichada.

Venezuela has three major Biogeographic regions: the Andean (4000–5000 m) and Coastal (2000–3000 m) cordilleras with montane and cloud forest on their slopes; and the Orinoco Llanos of semi-arid, seasonally-flooded savannah and shrub that covers the river's alluvial plain. The seasonal evergreen and semideciduous forests in the western llanos have been lost, replaced by cattle-ranching and agriculture. **Amazonian Moist Lowland, Submontane and Montane Forest** prevail south of the Orinoco.

> *Notable species:* in Colombia: the Cotton-top and White-footed tamarins (*Oedipomidas oedipus* and *O. leucopus*), the Colombian Squirrel Monkey (*Saimiri albigena*), three white fronted capuchins (*Cebus versicolor*, *C. cesarae* and *C. malitiosus*), four night monkeys (*Aotus lemurinus*, *A. griseimembra*, *A. brumbacki* and *A. jorgehernandezi*), the Ornate Titi (*Plecturocebus ornatus*), the Colombian Red Howler (*Alouatta seniculus*), the Variegated Spider Monkey (*Ateles hybridus*), the Brown Woolly Monkey (*Lagothrix lagothricha lugens*) and, in northern Venezuela, the Sierra de Perijá White-fronted Capuchin (*Cebus leucocephalus*), the Guianan Weeper Capuchin (*Cebus olivaceus*), the Ursine Red Howler (*Alouatta arctoidea*) and the Variegated Spider Monkey (*Ateles hybridus*). Trinidad has two endemics, the Trinidad White-fronted Capuchin (*Cebus trinitatis*) and the Trinidad Red Howler (*Alouatta seniculus insulanus*).

3. The foothills and montane forests of the Andes in southern Colombia, eastern Ecuador, Peru and Bolivia

The primate communities are similar to those of the moist tropical forests of Amazonia, despite occupying distinct habitats in terms of climate (above all, temperature and seasonality), topography and floristic communities in the **Submontane** and **Montane Cloud Forests** of the Andean valleys of the upper reaches of the major Amazonian rivers such as the Putumayo, Napo, Marañón, Huallaga and Ucayali. Large atelids, notably howler monkeys, spider monkeys and woolly monkeys, can be found at elevations up to 2800–3200 m.

> *Notable species:* Graells's Black-mantled Tamarin (*Leontocebus nigricollis graellsi*), the Andean Saddle-back Tamarin (*Leontocebus leucogenys*), the Andean Night Monkey (*Aotus miconax*), the San Martín Titi Monkey (*Plecturocebus oenanthe*), the Red Howler Monkey (*Alouatta seniculus*), the Black Spider Monkey (*Ateles chamek*), the White-bellied Spider Monkey (*Ateles belzebuth*), the Peruvian Yellow-tailed Woolly Monkey (*Lagothrix flavicauda*) and the Peruvian Woolly Monkey (*Lagothrix lagothricha tschudii*).

4. Amazonia

The Amazonas and Orinoco basins and the contiguous forests of Guyana, Suriname and French Guiana make up the world's largest rain forests and are collectively referred to as Amazonia. They include the Andean forests up to 500–1000 m, the ancient Pre-Cambrian shields of Guiana extending to the Orinoco in Venezuela, and Brazil south to the Cerrado in Central Brazil, as well as the alluvial plains of the Rio Amazonas/Solimões between them. The forests of the Guianas and Venezuela, although part of the Amazonian Hylaea, are not part of the Amazonian drainage basin since their rivers, the Orinoco in Venezuela and many rivers in Guyana, Suriname and French Guiana, flow north into the Atlantic. In eastern Brazil, the Rio Tocantins, fed by its tributary the Araguaia, flows directly into the Atlantic Ocean, although much of its drainage basin is also considered Amazonian rain forest. **Terra Firme Rain Forest** on plateaus above the flood level of the rivers is the most extensive of the forest types in Amazonia. It has a closed canopy at 25–35 m, with emergents up to 60 m tall. The plant biomass of these forests exceeds that of any other forest type. The extraordinary diversity of plant species is not spread uniformly throughout Amazonia. Some species are widespread, while others are local or regional endemics. Although physiognomically similar over immense areas, Amazonia is divided into eight distinct phytogeographic regions, based on their palaeohistory and present-day climate (Ducke & Black, 1953). The floristic communities are dependent on altitude, topography, soils, drainage and climatic seasonality. The *terra firme* rain forest is divided into **Dense Lowland Forest** (below 250 m elevation) with a canopy at 30 m, emergents reaching 60 m, and relatively open understories with few shrubs, on the Tertiary sediments between the major rivers, and **Hill Forest** (above 250 m), which has a more open and lower canopy and emergent trees up to about 35 m.

Amazonia has heavy rainfall and large areas of forest that are inundated. **Inundated Forests** are divided into those that are permanently, seasonally or periodically inundated, and also whether they are flooded by so-called (1) white water (turbid, brown, silt-laden, nutrient-rich waters of the Rio Amazonas and most of the rivers draining the Andes in western Amazonia); (2) black water (tea-coloured, low pH due to tannic acids, and nutrient-poor waters in most of the tributaries draining the western part of the Guiana Shield, notably the Rio Negro, along with some south-bank tributaries of the Rio Solimões draining predominantly sandy soils in central Amazonia such as the rios Jutaí, Tefé and Coarí; and (3) clear-water (clear, higher pH and relatively nutrient-poor of most of the rivers draining the Brazilian Shield and the eastern part of the Guiana Shield, notably the rios Aripuanã, Tapajós, Xingu, Tocantins and Araguaia). Permanently inundated, eutrophic (rich in nutrients) **Swamp Forests** are found along all water types. **Palm Swamps**, principally of *Mauritia* palms but also of *Euterpe* palms, grow on dystrophic soils (low in nutrients). Periodic flooding occurs twice a day in the **Mangrove Forests** of the Amazon and Orinoco deltas, as well as in forests along the lower Amazon up to 100 km from the delta where the river is backed up by incoming tides. The vegetation in these areas is called **Tidal Várzea**, *várzea* being the name for white-water flooded forest. **Floodplain Forests** develop along the upper reaches of the rivers that suffer periodic flash floods during heavy rains. Lastly, there are the seasonally flooded forests of the white-water rivers, **Seasonal Várzea**, and of the black- and clear-water rivers, known as **Seasonal Igapó**. *Igapó* and *várzea* have distinct floristic communities. They are flooded every year for 4–9 months, generally to depths of 2–5 m but, depending on the river catchment, sometimes to 14 m.

There are numerous areas of nutrient-deficient **White-sand Soils** on former river or coastal beaches. In Brazil they are called **Campinas**, small patches of savanna-like vegetation, and **Campinarana** or **Amazonian Caatinga**, larger areas of closed-canopy, low-lying forests, with shallow leaf litter and trees with small, scleromorphic leaves. The latter are particularly prevalent on the Guiana Shield in the Rio Negro basin north of the Rio Amazonas in Brazil and Venezuela, but also occur on the Brazilian Shield, south of the Amazonas, a notable example being the Serra do Cachimbo on the border of the Brazilian states of Mato Grosso and Pará.

Amazonian Savannas, open grasslands in more seasonal parts of the region, are similar to the Campo Cerrado (see below) of central Brazil. The largest is the Roraima–Rupununi savanna on the Brazil–Guyana border but there are smaller but nonetheless significant ones such as the Sipaliwini Savanna in Suriname and along the basin of the Rio Madeira. **Periodically Inundated Savannas** are found on the island of Marajó in the Amazon delta and also inland between the rios Xingu and Tapajós, for example.

Transition Forests form a belt between the high, closed canopy rainforest and areas of Cerrado or savanna (see below). They are lower, more seasonal in rainfall, have a more open canopy, and are often semideciduous. They include **Babaçú Palm Forest**, an open transitional forest dominated by *Orbygnia phalerata* palms along the borders of south-east Pará with the states of Maranhão and Tocantins; **Liana Forest** (*mata de cipó* or *cipoal*), which are open forests, rich in woody climbers, notable between the rios Tapajós, Xingu and Tocantins, but also in Roraima, Brazil, and in the southern parts of the Peruvian Amazon; and **Bamboo Forest**, with dense understorey bamboos (*taquara*) and giant bamboos that break the forest canopy over a large area stretching from southern Amazonian Peru across the state of Acre in Brazil and into the Pando region of Bolivia.

> *Notable species:* most of the Neotropical primates—73% (138/190) of species and 69% (150/218) of species plus subspecies—are partially or entirely Amazonian. Pygmy marmosets (*Cebuella*), the Black-crowned Dwarf Marmoset (*Callibella*), all but one of the 15 Amazonian marmosets (*Mico*), Goeldi's Monkey (*Callimico*), the saddle-back and black-mantled tamarins (*Leontocebus*), the moustached tamarins (*Tamarinus*), the tamarins of the Guiana Shield (*Saguinus*), the collared titis (*Cheracebus*) and the pitheciines—the sakis (*Pithecia*), bearded sakis (*Chiropotes*) and uacaris (*Cacajao*)—occur nowhere else.

5. The Brazilian Cerrado and Caatinga

The arid **Caatinga** of North-east Brazil is characterized by low arboreal deciduous sclerophytic scrubland in regions with less than 200–1000 mm of annual rainfall. It is divided into three principal vegetation types: **Arboreal Caatinga** consisting of dense tall scrub forest; **Shrubby Caatinga** dominated by dense spiny shrubs; and **Parkland Caatinga**, with open savanna-like xeric vegetation, scattered trees and numerous cactuses. There are similar arid regions in Central America, northern Colombia and northern Venezuela. A marked feature of the Caatinga is the presence of humid forests on the windward sides and tops of hills and mountains that result from orographic rainfall. In Brazil, they are called **Brejos**.

The **Cerrado** of the *planalto* region of Brazil has markedly seasonal annual rainfall of 1500–2000 mm. It is a complex of bush and forest savannah: **Cerradão**, low dense evergreen forest, 5–15-m tall with a closed but discontinuous canopy; **Campo Cerrado**, open savanna with intermittent, tortuous trees with thick fire-resistant bark, 2–5 m in height; and **Campo Limpo**, open grassland.

> *Notable species:* the White-tufted-ear Marmoset (*Callithrix jacchus*), the Black-tufted-ear Marmoset (*C. penicillata*), the Bearded Capuchin (*Sapajus libidinosus*), the Yellow-breasted Capuchin (*S. xanthosternos*), the Hooded Capuchin (*S. cay*), the Blond Titi (*Callicebus barbarabrownae*) and the Black and Gold Howler (*Alouatta caraya*). There are hybrid zones of *Callithrix penicillata* and *C. jacchus* in the north of the state of Bahia.

6. The Atlantic Forest

The Atlantic Forest (*sensu stricto*) includes the coastal **Moist Tropical Forests** in Brazil, from the state of Rio Grande do Norte through Paraíba, Pernambuco, Alagoas, Sergipe and Bahia in North-east Brazil, south to the Serra do Mar in the states of southern Espírito Santo, Rio de Janeiro, São Paulo, Paraná, Santa Catarina and north-east Rio Grande do Sul. *Sensu lato* it extends inland as **Semideciduous Forest** in the states of eastern Minas Gerais, São Paulo and northern Paraná, and as **Subtropical Semideciduous Alto Paraná Atlantic Forest** in the eastern parts of the states of Paraná, São Paulo and Santa Catarina in Brazil, in eastern Paraguay, and in the province of Misiones in north-eastern Argentina, as well as the **Subtropical Moist Mixed Paraná Pine** (*Araucaria angustifolia*) **Forest** of the montane plateaus of Paraná, Santa Catarina and northern Rio Grande do Sul, and the northern province of Misiones.

> *Notable species:* there are 24 species and subspecies in the Atlantic Forest—17 of them endemic. Endemics include the four species of lion tamarins (*Leontopithecus*), the two muriquis (*Brachyteles*), four marmosets (*Callithrix*), three titi monkeys (*Callicebus*), three robust capuchins (*Sapajus*) and the Brown Howler (*Alouatta guariba*). A few populations of the otherwise Amazonian Red-handed Howlers (*Alouatta belzebul*) survive in tiny forest fragments in the North-east, north of the Rio São Francisco, indicative of past connections with eastern Amazonia. Two marmosets, the Black-tufted-ear Marmoset, *Callithrix penicillata* (predominantly of the Cerrado), and the White-tufted-ear Marmoset, *C. jacchus*, native to the Atlantic Forest and Caatinga of North-east Brazil, have been introduced into the forests of South-east Brazil. They are widely invasive and due to hybridization and competition threaten the two endangered endemic marmosets, *Callithrix aurita* and *C. flaviceps*.

7. Southern Cone: northern Argentina, Paraguay and southern Bolivia

The vegetation in south-east Bolivia is the predominantly xeric deciduous forest known as the **Dry Chaco**. It extends into the plains of northern Paraguay west of the Río Paraguay. Progressively higher rainfall permits the appearance of humid semi-deciduous forests in southern Paraguay and northern Argentina, termed **Humid Chaco**. Besides occurring in southern Paraguay, a population of the Hooded Capuchin (*Sapajus cay*), isolated by the intervening Dry Chaco of Bolivia, is found in the southern Andes in Bolivia (departments of Chuquisaca and Tarija) and Argentina (provinces of Jujuy, Salta and Tucumán) in the **Subtropical Montane Forests of Las Yungas** along the eastern slopes of the Andes. We include in this region the northernmost basin of the upper Río Paraguai, the immense seasonally flooded alluvial plain (depression) known as the **Pantanal** in central western Brazil (states of Mato Grosso and Mato Grosso do Sul) and along the borders of eastern Bolivia and north-eastern Paraguay. It is a biogeographic meeting point for the floras and faunas of the Amazonian rain forest, Cerrado, the Atlantic Forest, the Chaco and the Chiquitano dry forest of Bolivia (Tomas *et al.*, 2022).

> *Notable species:* the Black-tailed marmoset (*Mico melanurus*), Azara's Night Monkey (*Aotus azarae azarae*), the Hooded Capuchin (*Sapajus cay*), the Southern Black-horned Capuchin (*Sapajus cucullatus*), the Pale Titi (*Plecturocebus pallescens*) and the Black and Gold Howler (*Alouatta caraya*). The northern Pantanal has records for the otherwise Amazonian Black Spider Monkey (*Ateles chamek*) and Mittermeier's Tapajós Saki (*Pithecia mittermeieri*).

All Neotropical primates are arboreal, that is, they live in three-dimensional habitats. In some localities in western Amazonia primate communities are among the richest in the world. They can consist of four species of callitrichids, two capuchins, a squirrel monkey, a night monkey, two titi monkeys, a saki and a bearded saki (or uacari), a howler monkey, a woolly monkey and a spider monkey—in all, 15 species. This is possible because each species differs with regard to the foods they exploit and where they obtain them. The 'where' is manifested in differences in diet associated with how they obtain, ingest and digest their food, and in their use of the different meso-ecosystems, i.e. the forest layers (vertical) and the subtle changes in floral and faunal communities on, for example, differing soils and drainages (horizontal) that affect the composition, dispersion (clumped or spread out) and abundance of the plant species upon which they depend. The vertical component of a tall tropical forest can be discerned by recording the amount of time the groups of each primate species spend in the different layers, from the **ground and shrub layer** to the **understorey** of small trees, bushes and lianas that extends to the branches of the big trees that make up the **main canopy** of the forest. The so-called '**emergents**' are the massive trees that habitually extend beyond the main canopy. Identification of the horizontal components is trickier, requiring an understanding of the composition of the floral communities and the type, dispersion and seasonality of the foods they offer to satisfy the particular dietary tendencies of each species.

Two terms frequently used to describe forests are **Primary Forest** and **Secondary Forest**. The names are useful for distinguishing, on the one hand, tall, seemingly pristine and little disturbed forests and, on the other, lower, denser, disturbed or degraded forests. In fact, they just describe two stages of succession in the slow-paced dynamics of ever-changing forests. A primary forest can arise from a stretch of riverbed left dry or the gradual drying out of an ox-bow lake after a change in the course of a river. The shrubs, small trees and palms in the early stages of succession are a primary (the first) forest. As the forest grows and ages, trees in tall forests die or are blown down leaving a gap in the canopy. Tree-fall gaps are then gradually filled by a secondary forest, which has its own sequence of successional stages. The secondary forest eventually fills the gap and left alone will achieve the stature and physiognomy of a 'primary forest'. A skilled botanist can tell by its floristic composition that it is secondary as it is dominated by tall trees that are characteristic of each stage of succession. Secondary forests arise in areas that have been disturbed or deforested in some way—either by tree falls, selective logging, patches that have been cleared for slash-and-burn farming or an encampment—and are subsequently reclaimed by the forest. The primary forest is a mosaic of secondary and primary forest patches. **Old Growth** or **Climax Forest** is perhaps a more precise name than 'primary'.

Threats

Non-human primates worldwide are threatened with extinction and more than 60% of all species and subspecies are now in the Critically Endangered, Endangered or Vulnerable categories of the Red List of Threatened Species of the International Union for Conservation of Nature (IUCN) (see below). All Neotropical monkeys are arboreal and most live in the tropical rain forests of this region. Unsurprisingly, the greatest threat they face is the destruction, fragmentation and degradation of their forests, although hunting is also a major concern for the larger species, especially the atelids. Hybridization and competition with invasive species threaten the Pied Tamarin (*Saguinus bicolor*) in Amazonia and, as mentioned, the marmosets of the southern Atlantic Forest (Tabarelli *et al.*, 2005; Oliveira & Grelle, 2015; Moraes *et al.*, 2019). In the Atlantic Forest, yellow fever outbreaks are also a cause of high mortality, affecting most of all the howler monkeys. Wildlife trafficking was once a serious threat (1960s to 1970s) but, albeit still ongoing, is now much diminished.

Forest destruction is total in areas that have been clear-cut for agriculture and cattle pasture, flooded by reservoirs or subject to mining. Fragmentation creates the checkerboard of forest patches of different sizes remaining in a mosaic of different land uses, extensively occupied by open spaces and barriers such as urban areas, roads, railways and electricity transmission lines. These open spaces isolate the primate communities and only the larger forest patches provide conditions for their long-term survival. Degradation is provoked by wind and other edge effects, the use of forest resources, logging (intensive and selective), oil and gas exploration, and rural colonization that results in hunting, slash-and-burn agriculture and pollution. Fragmentation facilitates hunting and access to the forest, and colonization and urbanization bring the dangers of electrocution, roadkill, zoonotic disease transmission, and attacks by domestic dogs, and, to a variable extent, changes in microclimates that alter the floral and faunal composition of the forest patches. The loss of species in each fragment subtly but completely changes predator-prey relations and the composition and dynamics of forest animal and plant communities.

The impact of subsistence hunting can be huge in many areas, especially for the medium to larger-sized genera such as *Cebus*, *Sapajus*, *Chiropotes*, *Cacajao*, *Alouatta*, *Lagothrix*, *Ateles* and *Brachyteles*. The Northern Muriqui (*Brachyteles hypoxanthus*) was probably extirpated from southern Bahia and large areas of its original range by hunting. Over the past 50+ years, we have visited areas of completely intact old-growth forest in Amazonia where, for example, *Ateles* and *Lagothrix* species—which should be there—are now entirely absent or very rare. This 'empty forest syndrome' is prevalent in many areas and is a major concern. Smaller species tend to be less persecuted and are often common in secondary forests close to towns and villages. In some areas, however, callitrichids are selectively hunted. The Matis people of the Rio Javarí basin, for example, hunt tamarins to use their teeth in necklaces and bracelets that are status symbols.

The worldwide assault on tropical forests, which began in earnest in the 1970s, initially targeted timber. It also promoted rural colonization and, consequently, agriculture and cattle ranching. Subsequently, industrial and artisanal mining arrived, along with the flooding of forests for hydroelectric energy, all of which severely damage fluvial and hydric dynamics and, in the process, pollute many lakes and rivers. Today, the number-one threat is unquestionably forest clearance for large-scale agro-industry for soy, which has fragmented and destroyed more than 500 thousand km^2 forest in the 'arc of deforestation' that is rapidly moving north from the northern part of the Brazilian state of Mato Grosso in the south of Pará, and westward through the states of Rondônia and Acre. The forests of the Brazilian Amazonian state of Maranhão have been almost entirely destroyed. The Andean countries, Colombia, Ecuador and Peru, are likewise suffering the constant widespread destruction of their forests in both montane and lowland Amazonia. Oil palm plantations are also starting to have an impact, although as yet nowhere near the devastation that they have caused in the forests of South-east Asia. Colombia is today the world's third-highest producer of palm oil, and this industry is already beginning to have a huge impact on its forests. Other major threats come from large-scale industrial mining for such as bauxite and iron, *Eucalyptus* and pine plantations for the pulp and paper industry and, notably in Ecuador, petroleum exploration. All these activities permanently eliminate and pollute vast expanses of forest and their rivers. Climate change is a growing threat for Neotropical primates due to habitat loss and, with the changes in temperature and rainfall (climatic seasonality) that it provokes, is increasingly affecting forest faunal and floral communities, i.e. the availability, quality, abundance and dispersion of food sources. It may also make them more vulnerable to disease. This year, 2024, has seen horrific die-offs of Mexican mantled howlers due to heatstroke.

Conservation status

The globally recognized database on the conservation status of the world's animals, plants and fungi around the world is that of the Species Survival Commission (SSC) of the International Union for Conservation of Nature (IUCN). The SSC is composed of more than 9500 volunteer experts in 186 Specialist Groups and task forces, the majority of which are taxonomic, focussing on specific animal and plant groups. Each group contributes to the **IUCN Red List of Threatened Species**, a systematic, documented assessment of the conservation status of the species in their remit. Using well-established criteria, each species and (for primates at least) subspecies is assigned to one of nine categories as follows: **Critically Endangered** CR – the species is facing an extremely high risk of extinction; **Endangered** EN – the species is facing a very high risk of extinction; **Vulnerable** VU – the species is facing a high risk of extinction; **Near threatened** NT – the species is close to being at risk of extinction; **Least concern** LC – the species is not currently at risk of extinction; **Data Deficient** DD – there is insufficient information to assess the conservation status of the species; and **Not Evaluated** NE – the conservation status of the species has yet to be assessed. No Neotropical primate has yet been categorized as **Extinct in the Wild (EW)** or **Extinct (EX)**. These assessments are based on demographic aspects, the number of populations, their size and the extent of their fragmentation, estimated rates of the decline and loss of habitat, the species' geographical range (Extent of Occurrence), and an understanding of where the species is actually found (Area of Occupancy), in addition to an assessment of the threats each species is facing. The IUCN SSC Primate Specialist Group has been very active in the Red List process for the past 50 years and has conducted multiple workshops to determine the status of primates around the world.

The most recent assessments have shown that 214 species and subspecies (43%) are endangered (CR+EN) and, adding the category of Vulnerable (VU), the number that are threatened with extinction (CR+EN+VU) rises to 468 (64%). This makes primates one of the two most threatened larger groups of vertebrates in the world, the other being the turtles. Considering just the Neotropical primates, 53 (24%) are endangered (CR+EN) and 95 (44%) are threatened (CR+EN+VU). Table 3 shows the number of primates, the number of endemic species and subspecies, and the number and percentage of threatened primates in each country.

As can be seen in Table 4, the families differ in the extent to which they are threatened. Habitat loss and fragmentation, both ongoing and potential, are universally regarded as the main threat to all Neotropical primates, often with aggravating factors such as hunting (medium-to-larger-sized species) and the pet trade, which, despite concentrating on smaller species, does often also target medium-to-larger-sized species.

Callitrichids are threatened by forest loss in the Atlantic Forest and in central and northern Colombia, and a few species with small ranges are also at risk in Amazonia. Ongoing forest destruction and fragmentation is the main threat and the principal reason for the Critically Endangered status of the Cotton-top Tamarin (*Oedipomidas oedipus*) of northern Colombia and the Endangered status of the four lion tamarins (*Leontopithecus*) in the Atlantic Forest of eastern Brazil. In past decades, the trade for biomedical purposes was an aggravating factor for the former and the pet trade for the latter. Two marmosets in South-east Brazil, the Buffy-tufted-ear Marmoset (*Callithrix aurita*) and the Buffy-headed Marmoset (*C. flaviceps*), are not only losing their forests, but also suffering from hybridization with the Black-tufted-ear Marmoset (*C. penicillata*) and the White-tufted-ear Marmoset (*C. jacchus*). Many callitrichids have restricted ranges making them even more susceptible to the destruction of their forests. This is the case of *C. flaviceps* and particularly the Pied Tamarin (*Saguinus bicolor*) in Central Amazonia, which has a tiny geographic range in and around the sprawling city of Manaus, capital of the state of Amazonas. The Pied Tamarin is also facing the threat of hybridization and being replaced by the Golden-handed Tamarin (*S. midas*). The recently described Schneider's Marmoset (*Mico schneideri*) is Endangered because its small geographical range is in the industrial agriculture frontier (principally for soybeans) in the north of the state of Mato Grosso.

Squirrel monkeys are small, so hunting is less of an issue, except for *Saimiri oerstedii* in some areas in Central America where they are targeted because of the damage they cause to crops. Although most have large well-protected distributions in Amazonia, three do not: the two subspecies of the Central American Squirrel Monkey (*Saimiri oerstedii*) and the Black-headed Squirrel Monkey (*S. vanzolinii*), which are now Endangered. The latter occurs in the Central Amazon in a small interfluvial *várzea* that is highly susceptible to climate change and evidence is emerging that two other squirrel monkey species are invading its restricted range.

The larger capuchin monkeys are generalist omnivores and flexible in their behaviour, so are adaptable to areas with human disturbance. Nonetheless, they are hunted for food, sometimes heavily, and suffer retaliation due to crop raiding. Three robust capuchins, the Crested Capuchin (*Sapajus robustus*), Yellow-breasted Capuchin (*S. xanthosternos*) and Blond Capuchin (*S. flavius*), of the Atlantic Forest and Caatinga are now Endangered as a result of hunting and the loss and extreme fragmentation of their forests. The gracile capuchins (*Cebus*) are generally widespread, although one species, *Cebus kaapori*, is now severely reduced in numbers and Critically Endangered due to hunting and the extreme destruction and fragmentation of its forests in the eastern Amazonian states of Pará and Maranhão. Three restricted-range gracile capuchins, the Varied White-fronted Capuchin (*Cebus versicolor*), the Río Cesar White-fronted Capuchin (*C. cesarae*) and the Santa Marta White-fronted Capuchin (*C. malitiosus*), are in similar situations in northern Colombia and Endangered. The Ecuadorian White-fronted Capuchin (*C. aequatorialis*) in eastern Ecuador, west of the Andean Cordillera, and the Trinidad White-fronted Capuchin (*C. trinitatis*) on the island of Trinidad are both Critically Endangered. The latter, along with the native Trinidad red howler, *Alouatta seniculus insulanus*, is threatened by alien species introduced by the pet trade from Venezuela. Hundreds of introduced robust capuchins (*Sapajus*), Venezuelan white-fronted capuchins (*Cebus*), and squirrel monkeys (*Saimiri*) now reside on the western peninsula in the Chaguaramas Macqueripe area, and are displacing the local Trinidad primates through competition for food and habitat, and threaten particularly the native capuchins through hybridization (J. Seyjagat, in litt., 2024).

Until 1983, the night monkeys, *Aotus*, the only nocturnal monkeys, were considered to consist of a single species, *A. trivirgatus*. Today, 13 species and subspecies are recognised. They occur from Panama and Venezuela south to northern Argentina. Elusive and small (about 1 kg), they are not generally hunted for food or as pets, but all are affected by forest loss, fragmentation and degradation. The Andean Night Monkey (*Aotus miconax*), an Endangered Peruvian endemic, has a restricted range in pre-montane and montane forests in the departments of Amazonas, Huánuco, La Libertad, Loreto and San Martín. Hernández-Camacho's Night Monkey (*A. jorgehernandezi*), described in 2007, is known only from a single specimen from the western slopes of the Andes in the regions of Quindío and Riseralda, Colombia. Night monkeys have been captured for biomedical research since the 1950s and are particularly valued for malarial research. Among those threatened by this trade (illegal and legal) is Ma's Night Monkey (*A. nancymai*). It is categorized as Vulnerable on the IUCN Red List and found only in restricted range in western Brazil south of the Rio Amazonas/Solimões. Exported from Peru and Colombia to the USA, today it accounts for about 40% of the biomedical trade of primates from South America, much of which is illegal.

Thirty-five of the 63 species of the Family Pitheciidae are titi monkeys, which are rarely hunted. Eight titi monkeys are endangered due to the generalized destruction of their forests in their geographically restricted ranges: the Río Beni Titi (*Plecturocebus modestus*) and the Olalla Brothers' Titi (*P. olallae*) in northern Bolivia; the San Martín Titi (*P. oenanthe*) in northern Peru; the Caquetá Titi (*P. caquetensis*) in southern Colombia; Groves's Titi (*P. grovesi*) trapped in the 'arc of deforestation' in the north of the state of Mato Grosso, Brazil, where much of its range has been lost to soybean plantations or flooded by a dam on the Rio Xingu; two Atlantic Forest species, Coimbra-Filho's Titi (*Callicebus coimbrai*) and the Masked Titi (*C. personatus*); and the Blond Titi (*C. barbarabrownae*), endemic to the Caatinga.

The 16 sakis of the genus *Pithecia* are a little larger but generally not hunted, although in some areas they are a preferred species, even though difficult to find. Their ranges are larger and none are considered endangered, although six are categorized as Data Deficient on the IUCN Red List. The 13 bearded sakis and uacaris are hunted and six are Vulnerable as a result. The Black Saki (*Chiropotes satanas*) occurs in the same devastated region as the Ka'apor Capuchin (*Cebus kaapori*) and is Endangered.

The largest Neotropical monkeys are the 40 species and subspecies of howler monkeys, spider monkeys, woolly monkeys and muriquis belonging to the Family Atelidae. All are hunted for food but, unlike in Africa and Asia where primate meat is commercialized, they are hunted largely for subsistence. The howler monkeys, *Alouatta*, are distinct from the atelins in certain demographic and behavioural aspects that make their populations more resilient to the impact of hunting. Owing to their folivory (much of their diet consists of leaves), howlers have smaller home ranges, which means that population densities are higher. Males reach puberty earlier and females begin reproducing at an earlier age and have shorter gestation periods and shorter intervals between births than the atelins. The more frugivorous atelins (especially spider monkeys) have larger group home ranges (lower population densities) and slower life histories. Compared to the howlers, males take longer to reach puberty, females begin breeding later and have longer gestation periods, and the intervals between births are longer than in howlers. Females of some howler monkey species may have offspring twice in three years, whereas the spider monkeys, woolly monkeys and muriquis will have just one. The smaller group home ranges of the howlers also means that they can maintain populations in small forest fragments that are otherwise insufficient for the needs of the atelins. Half of the atelid species and subspecies are Endangered and three-quarters are threatened. Five of the six howler monkeys from Central America and Mexico are Endangered, of which one, the Mexican Mantled Howler, *Alouatta palliata mexicana*, has the lowest genetic diversity of any primate. The sixth, the Ecuadorian Mantled Howler, *Alouatta palliata aequatorialis*, is categorized as Vulnerable. Of the 14 South American howlers, six are threatened. The Maranhão Red-handed Howler (*A. ululata*) is Endangered due to extensive forest loss and hunting. The genus *Alouatta* is also highly susceptible to the African yellow fever virus and recent outbreaks have drastically reduced and even locally extirpated populations of the Brown Howler (*A. guariba*) in the Atlantic Forest, as well as affecting many other threatened primate species. All 13 species and subspecies of spider monkeys (*Ateles*) in Mexico and Central and South America are now categorized as threatened—11 are Endangered or Critically Endangered. Finally, the two muriqui species, of Brazil's Atlantic Forest, the Southern Muriqui (*Brachyteles arachnoides*) and the Northern Muriqui (*B. hypoxanthus*), are Critically Endangered.

Conservation action plans have been drawn up in recent decades that review the species' threatened status and establish coordinated measures for the conservation of primate species and communities worldwide. The first such plan for a Neotropical primate was for the lion tamarins, *Leontopithecus*, *Saving the Lion Marmosets* (Bridgewater, 1972). In 1977, the IUCN SSC Primate Specialist Group published *A Global Strategy for Primate Conservation*, that reviewed the status of primates worldwide and the conservation measures needed (Mittermeier, 1977). Subsequent plans were drawn up by the IUCN SSC Conservation Planning Specialist Group, which had established a methodology for species conservation based on Population and Habitat Viability Analyses (PHVA) and for species groups based on Conservation Assessment and Management Plans (CAMP). These have been used to inform conservation action planning for a number of Neotropical primates. The recognition of biodiversity conservation as a global priority has resulted in several national primate conservation action plans. In Latin America, government implementation of national biodiversity conservation commitments has been a turning point for primate conservation planning as many countries—Brazil, Mexico, Peru, Ecuador and Argentina—have organized national action planning efforts for the conservation of threatened primates or set up officially recognized plans developed by national primatological societies. This has pushed forward primate conservation as these efforts are officially stated and implemented as a matter of public policy, which brings with it avenues for funding, the inclusion of standardized protocols in environmental licensing processes, established official population management programmes, and defined priority areas for species conservation (Strier *et al.*, 2021; Reuter *et al.*, 2022a).

To promote public awareness of the critical situation of primates around the world, in 2000 the IUCN SSC Primate Specialist Group (PSG), together with Conservation International, drew up a list of the *World's 25 Most Endangered Primates* and produced a brief report containing short accounts of each of the 25 species identified and the threats to their survival (Mittermeier *et al.*, 2000). The aim was to draw attention worldwide to Critically Endangered primates that were lacking the conservation measures urgently needed for their survival, targeting not merely supportive public opinion but most especially alerting governments, donors and funding sources, researchers and conservation NGOs to the need for action. The list, launched in 2000, received exceptional coverage in a media environment already saturated with millennial news and even earned a two-page spread in *Time* magazine. Its success stimulated the PSG to repeat the exercise but this time consulting primate researchers and conservationists when deciding which species should be included. An open meeting was held during the 19th Congress of the International Primatological Society (IPS) in Beijing, China, in which primatologists contributed information fresh from the field. The second updated list and 2002–2004 report were released in Johannesburg, South Africa, on 2 September 2002. The revision of the 2000 list culminated in the official endorsement by the IPS of the 'Top 25', which has since become a joint endeavour at their biennial congresses (Reuter *et al.*, 2022b). Twelve biennial lists have been drawn up and published since 2000 profiling the plights of 103 species, 24 of them Neotropical.

Primate watching and primate tourism

As with our previous field guides, this illustrated checklist aims to continue to promote the hobby/sport of primate-watching and its associated activity, primate life-listing. Needless to say, the idea for this derives from birdwatching, one of the most popular hobbies throughout North America, Europe, Australia, South Africa and increasingly elsewhere in the world, especially in the Neotropical region. Birdwatching has been with us for a long time and its popularity is still growing. It has benefited from an ever-increasing number of guidebooks that cover the entire planet and over the past 15 years by the availability of new sophisticated smartphone applications for bird identification that provide information on the birds' appearance and calls. The most striking example is the Merlin Bird ID phone application (*merlin. allaboutbirds.org*), released for free by the Cornell Lab of Ornithology (Cornell University, New York), that has about 700,000 active users per month—and counting. Huge progress has resulted from websites connecting birders around the world and from global bird databases such as eBird (*ebird.org*), hosted by the Cornell Lab, and regional and national databases such as the Euro Bird Portal (*eurobirdportal.org*) where birders upload their observations. In recent years, eBird, which is available in 48 languages, has recorded over 100 million bird observations annually and in May 2021 reached the milestone of one billion submitted bird observations. All of this has been good for conservation, stimulating awareness of and love for birds, and provided many ecotourism-based economic opportunities for communities living in or near bird habitats throughout the world. The passion for birds has become a multi-billion-dollar industry, with at least some of the benefits accruing to the bird-rich countries of the tropics.

In the USA alone, US$41 billion is spent annually on birdwatching activities (including garden birding, feeders, seeds and other foods, binoculars, cameras and other equipment). Of this, US$17 billion is spent on birdwatching travel to a wide variety of destinations, both domestic and international (Withrow, 2019). As another example, a recent study using eBird data looked at trends in Alaska birdwatching and applied existing information from the Alaska Visitor Statistics Program to estimate visitor expenditure and the impact of spending on Alaska's regional economy: in 2016 it was calculated that nearly 300,000 birdwatchers visited Alaska and spent US$378 million, which supported approximately 4000 jobs (Schwoerer and Dawson, 2022).

Likewise, the economic opportunities for tropical countries such as Brazil, Colombia, Peru, Ecuador, Indonesia, Kenya, Tanzania and many others, all with great bird diversities, is also very significant. The National Audubon Society estimated that 150,000 birdwatchers will visit Colombia from the United States during the decade 2017–2027, thereby generating US$47 million annually and sustaining 7500 new jobs (Ocampo-Peñuela & Winton, 2017). These authors indicate, however, that these numbers could even be an underestimate if Colombia can emulate the recent surge in birdwatching tourism in neighbouring Peru. Demand for birdwatching tourism appears to be sustainable, as the global market is already very large, with 46 million birdwatchers in the USA alone; moreover, research shows that this hobby is gaining popularity (Cordell and Herbert, 2002).

The second author of this book (Russell Mittermeier) was inspired by the success and impact of birdwatching and bird life-listing to launch, nearly 30 years ago, primate-watching and primate life-listing as a formally recognized activity (e.g., Coniff, 2007). There are in fact quite a few primate-watchers already, and some of us have been active for as long as five decades. Nevertheless, we have not yet given ourselves a name, we do not have a website, and we do not have listing software. What is more, we have not yet distinguished ourselves as a unique group, clan or tribe within the global community of nature-lovers. In addition, we (with a handful of exceptions) have not yet begun to compete with one another—an essential ingredient in any hobby or sport. By comparison with what exists for birds, we have relatively little in the way of published material for identifying primates, including country or regional field guides and other visual and auditory aids.

Fortunately, this is changing. We first tried to stimulate lemur-watching in 1994 with the first edition of a field guide to the lemurs of Madagascar (Mittermeier *et al.*, 1994), which since then has run to four more editions (Mittermeier *et al.*, 2006, 2010, 2014, 2023), and the *Primates of West Africa: A Field Guide and Natural History* (Oates, 2011). The first such guide for the Neotropics was for Colombia, published in 2003 in Spanish and in 2004 in English (Defler, 2003, 2004). A number of other authors have produced useful primate guides, including, for example, for Central Africa (Gautier-Hion *et al.*, 1999), Asia (Beauséjour *et al.*, 2021), southern Asia (Walker & Molur, 2007), south-east Asia (Shepherd & Shepherd, 2017), Indonesia (Supriatna, 2019), Brazil (Auricchio, 1995; dos Reis *et al.*, 2015), Colombia (Bennett, 2003), Ecuador (de la Torre, 2000) and the Guianas (Boinski, 2002; de Thoisy & Dewynter, 2004). In addition, primate information of variable quality can also be found in a number of other regional or national guidebooks on mammals, for example for Africa (Kingdon, 2015), China (Smith & Xie, 2008; Copete, 2020), southern Asia (Elliott & Martínez-Vilalta, 2020), South-east Asia (Francis, 2008), Borneo (Phillipps & Phillipps, 2018), Vietnam (Groves *et al.*, 2020), Ecuador (Tirira, 2007), Honduras (Marineros & Gallegos, 1998), Venezuela (Linares, 1998) and Central America and south-east Mexico (Reid, 1997).

Russell Mittermeier and Anthony Rylands also launched a series of Tropical Pocket Guides in 2004, first with Conservation International and now with Re:wild. They are small convenient, folding guides to identify animals from particular regions. Thirty-three have now been published, 28 of them on primates, including 13 for Neotropical primates.

Covers of 12 fold-out pocket identification guides for Neotropical primates produced by or in collaboration with the IUCN Species Survival Commission Primate Specialist Group (PSG).

Recently, we joined forces with *mammalwatching.com*, the first and today preeminent website dedicated to mammal-watching, which receives thousands of visitors every week. This site contains well over a thousand trip reports from the growing mammal-watching community worldwide and also hosts an active community discussion forum where subscribers can share information or seek advice, with an average of 10 new posts each week. The site also has information on mammal taxonomy, field guides and equipment. The mammal-watching podcast is now three-years old and has a loyal following who listen to interviews with prominent mammalogists, conservationists and mammal-watchers. We are supporting the development of a separate section on primates, including league tables where people can showcase their primate lists, which will facilitate listing by our community of primate-watchers. Although we are still way behind the birders regarding the number of quality publications that help us identify what we are seeing and still do not have primate-specific software or a website, we continue to improve the materials that we have available.

Why bother? Well, first of all, primate-watching and primate life-listing are fun. Those of us who are as passionate about these animals as birders are about their species really enjoy seeing monkeys, apes, lemurs, lorises, galagos, pottos and tarsiers in their natural environments and want more people to get excited about them. But it is about more than just entertainment. Our main objective is to stimulate awareness of primates through these activities. Second, primates are found mainly in tropical rain forests where they are generally the most visible forest mammals. As such, they have been—and continue to be—excellent flagships for these dwindling habitats and have contributed greatly to tropical rain forest conservation over the past 40–50 years. Furthermore, we need more primate-based ecotourism to provide the communities living in close proximity to the habitats in which primates live with economic alternatives. These communities need to benefit economically from the presence of primate populations if they are to play major roles in conserving them. To ensure that this happens, tourists, scientists and primate watchers need to go and see these creatures in their natural environments, interact with the communities upon whose survival they ultimately depend, share our excitement and enthusiasm, and, most importantly, make a contribution to the local economy. In many places, this may be the only effective tool at our disposal to ensure the survival of Critically Endangered and Endangered primates and it needs to happen now.

Indeed, some primate ecotourism already exists. In Central Africa, Mountain Gorilla tourism has been working for more than 40 years and is an excellent model to follow. Moreover, many new primate sites are being developed every year, including for other gorilla species and subspecies in Central Africa, chimpanzees in various countries, and orangutans in parts of Sumatra and Borneo. China has set up several sites for seeing the Golden Monkey and other snub-nosed monkey species (*Rhinopithecus*). Many macaque and langur species are easily seen at sacred sites and even in many urban areas in China, India, Bangladesh, Nepal and South-east Asian countries, as well as increasingly in natural forests. Lemur-watching in Madagascar is on a rapid upswing and more and more Neotropical monkeys can be seen in a wide variety of parks and reserves in Mexico, Central America and South America, something that we hope to stimulate further with the publication of this book.

Unfortunately, primate ecotourism has not always been conducted as well or as carefully as we might like and needs to be improved wherever the quality is poor or even detrimental to primates or the local human communities (see Starin, 2009). However, we need to recognize that it is a powerful way of promoting the conservation of tropical forests, the well-being of local communities, the economies of the countries where primates occur, and, of course, the survival of the primates themselves. We need, therefore, to make sure it is conducted in the most appropriate manner possible. Our IUCN SSC Primate Specialist Group has published best-practice guidelines for responsible primate ecotourism, notably for the great apes for which tourism is already well-established (e.g., Macfie & Williamson, 2010). Recently, the Group's Section on Human-Primate Interactions has published important indications regarding photographing primates (Waters *et al.*, 2021), and drawn up best-practice guidelines for primate watchers (Waters *et al.*, 2023a) and for the wildlife tourism industry in general when primates are involved (2023b). We have just started to scratch the surface of the potential that exists for primate-watching and still need to highlight at a much higher level the economic and conservation benefits that it can provide.

To further stimulate interest in Neotropical primates and to launch people on their primate-watching and primate life-listing careers, we provide here not just a comprehensive guide but also a checklist of species of this region (see page 126) to jumpstart lists for those travelling in the Neotropical region. We wish you success and hope that you will join us in this new and exciting activity and become another member of what will soon be a large worldwide community of primate-watchers.

The main purpose of this checklist is to add to the growing body of resources available for facilitating and stimulating primate ecotourism and, more specifically, primate-watching and primate life-listing. We have already produced a field guide to the primates of Colombia (Defler, 2003, 2004), as well as a series of pocket guides to the Atlantic Forest region of Brazil, the Guianas, Colombia, Ecuador and three Brazilian Amazonian states, Amapá, Rondônia and São Paulo. A number of country-specific primate books also exist, as well as several excellent guides to the mammals of the Neotropics (Emmons, 1997; Reid, 1967; Gardner, 2007; Patton et al., 2015). However, to date no guide has ever been published specifically on the primate fauna of this entire region; this book now fills this gap. We are also in the process of preparing a much larger and more detailed guide to Neotropical primates along the lines of our field guide, *Lemurs of Madagascar*, now in its fifth edition (Mittermeier et al., 2023).

We believe that primate ecotourism can make a major contribution to protecting these animals in the long term and in engaging local communities as partners in this endeavour, which will benefit local and regional economies and create a real groundswell of interest in and commitment to conservation. We have seen this work in many parts of the world, including this region, and we believe that the potential in the Neotropics is enormous. We hope that you enjoy this book, that you will use it in the field, and that you come to have the same great enthusiasm for these wonderful animals as all of us do.

References

Auricchio, P. 1995. *Primatas do Brasil*. Terra Brasilis, São Paulo.
Ayres, M.C. & Prance, G.T. 2013. On the distribution of pitheciine monkeys and Lecythidaceae trees in Amazonia. In: *Evolutionary Biology and Conservation of Titis, Sakis and Uacaris*, L.M. Veiga, A.A. Barnett, S.F. Ferrari & M.A. Norconk (eds.), pp.127–139. Cambridge University Press, Cambridge, UK.
Barnett, A.A., Bowler, M., Bezerra, B.M. & Defler, T.R. 2013. Ecology and behavior of uacaris (genus *Cacajao*). In: *Evolutionary Biology and Conservation of Titis, Sakis and Uacaris*, L.M. Veiga, A.A. Barnett, S.F. Ferrari & M.A. Norconk (eds.), pp.151–172. Cambridge University Press, Cambridge, UK.
Barrett, B.J., Monteza-Moreno, C.M., Dogandžić, T., Zwyns, N., Ibáñez, A. & Crofoot, M.C. 2018. Habitual stone-tool-aided extractive foraging in white-faced capuchins, *Cebus capucinus*. *R. Soc. Open Sci.* **5**: 181002.
Beauséjour, S., Rylands, A.B. & Mittermeier, R.A. 2021. *All Asian Primates*. Re:wild, Austin, TX, and Lynx Edicions, Barcelona.
Beck, R.M.D., de Vries, D., Janiak, M.C., Goodhead, I.B. & Boubli, J.P. 2023. Total evidence phylogeny of platyrrhine primates and a comparison of undated and tip-dating approaches. *J. Hum. Evol.* **174**: 103293.
Bennett, S.E. 2003. *Los Micos de Colombia*. Instituto de Investigación de Recursos Biológicos Alexander von Humboldt, Fundación Tropenbos Colombia, Bogotá.
Bezanson, M., Cortés-Ortiz, L., Bicca-Marques, J.C., Boonratana, R., Carvalho, S., Cords, M., de la Torre, S., Hobaiter, C., Humle, T., Izar, P., Lynch, J.W., Matsuzawa, T., Setchell, J.M., Zikusoka, G.K. & Strier, K.B. 2024. Words matter in primatology. *Primates* **65**: 33-39.
Bicca-Marques, J.C. & Heymann, E.W. 2013. Ecology and behavior of titi monkeys (genus *Callicebus*). In: *Evolutionary Biology and Conservation of Titis, Sakis and Uacaris*, L.M. Veiga, A.A. Barnett, S.F. Ferrari & M.A. Norconk (eds.), pp.196–207. Cambridge University Press, Cambridge, UK.
Boinski, S. 2002. *De Apen van Suriname / The Monkeys of Suriname*. STINASU, Paramaribo.
Bond, M., Tejedor, M.F., Campbell K.E. Jr., Chornogubsky L., Novo N. & Goin F. 2015. Eocene primates of South America and the African origins of New World Monkeys. *Nature* **520**: 538–541.
Brcko, I. C., Carneiro, J., Ruiz-García, M., Boubli, J. P., Silva-Júnior, J. S., Farias, I., Hrbek, T., Schneider, H. & Sampaio, I. 2022. Phylogenetics and an updated taxonomic status of the tamarins (Callitrichinae, Cebidae). *Mol. Phylogenet. Evol.* **173**: 10754.
Bridgwater, D.D. (ed.) 1972. *Saving the Lion Marmoset*; The Wild Animal Propagation Trust Conference on the Golden Marmoset, National Zoological Park, Washington, DC.
Byrne, H.M., Rylands, A.B., Carneiro, J., Lynch Alfaro, J.W., Bertuol, F., Silva, M.N.F. da, Messias, M., Groves, C.P., Mittermeier, R.A., Farias, I., Hrbek, T., Schneider, H., Sampaio, I. & Boubli, J.P. 2016. Phylogenetic relationships of the New World titi monkeys (*Callicebus*): first appraisal of taxonomy based on molecular evidence. *Front. Zool.* **13**: 10. DOI 10.1186/s12983016-0142-4.
Byrne, H.M., Costa-Araújo, R., Farias, I.P., Silva, M.N.F. da, Messias, M., Hrbek, T. & Boubli, J.P. 2021. Uncertainty regarding species delimitation, geographic distribution, and the evolutionary history of south-central Amazonian titi monkey species (*Plecturocebus*, Pitheciidae). *Int. J. Primatol.* **45**: 12-34.
Campbell, C.J. (ed.) 2008. *Spider Monkeys: Behaviour, Ecology and Evolution of the genus* Ateles. Cambridge University Press, Cambridge, UK.
Campbell, D.G. & Hammond, H.D. (eds.) 1989. *Floristic Inventory of Tropical Countries: The Status of Plant Systematics, Collections, and Vegetation, plus Recommendations for the Future*. New York Botanical Garden, New York.
Cartelle, C. & Hartwig, W.C. 1996. A new extinct primate among the Pleistocene megafauna of Bahia, Brazil. *Proc. Natl. Acad. Sci. U.S.A.*, **93(13)**: 6405-6409.
Carvalho, C.T. de. 1957. Nova subespécie de saguim da Amazônia. *Rev. Brasil. Biol.* **17(2)**: 219–222.
Coimbra-Filho, A.F. & Mittermeier, R.A. 1972. Taxonomy of the genus *Leontopithecus* Lesson 1840. In: *Saving the Lion Marmoset*, D. D. Bridgwater (ed.), pp.7–22. Wild Animal Propagation Trust, Wheeling, WV.

Conniff, R. 2007. Primate watching is the new birding. *Audubon Magazine*. Available online: www.audubon.org/magazine/july-august-2007/primate-watching-newbirding.

Cooke, S.B., Rosenberger, A.L. & Turvey, S. 2011. An extinct monkey from Haiti and the origins of the Greater Antillean primates. *Proc. Natl. Acad. Sci. U.S.A.* **108(7)**: 2699–2704.

Cooke, S.B., Gladman, J.T., Halenar, J.B., Klukkert, Z.S. & Rosenberger, A.L. 2016. The paleobiology of the recently extinct platyrrhines of Brazil and the Caribbean. In: *Phylogeny, Molecular Population Genetics, Evolutionary Biology, and Conservation of the Neotropical Primates*, M. Ruiz-Garcia and J.M. Shostell (eds.), pp.41–90. Nova Science, New York.

Cooke, S.B., Mychajliw, A.M., Southon, J. & McPhee, R.D.E. 2017. The extinction of *Xenothrix mcgregori*, Jamaica's last monkey. *J. Mammal.* **98(4)**: 937–949.

Copete, J.L. 2020. *Mammals of China*. Lynx Edicions, Barcelona.

Cordell, H.K. & Herbert, N.G. 2002. The popularity of birding is still growing. *Birding* **34(6)**: 54–61.

Defler, T.R. 1999. Locomotion and posture in *Lagothrix lagotricha*. *Folia Primatol.* **70(6)**: 313–327.

Defler, T.R. 2003. *Primates de Colombia*. Conservation International Tropical Field Guide Series, Washington, DC.

Defler, T.R. 2004. *Primates of Colombia*. Conservation International Tropical Field Guide Series, Washington, DC.

Defler, T.R. & Stevenson, P.R. (eds.) 2014. *The Woolly Monkey: Behavior, Ecology, Systematics, and Captive Research*. Springer, New York.

de la Torre, S. 2000. *Primates de la Amazonía del Ecuador / Primates of Amazonian Ecuador*. SIMBIOE, Quito.

de Thoisy, B. & Dewynter, M. 2004. *Les Primates de Guyane*. Collection Nature, Guyanaise, Sepanguy, Cayenne.

Dias, P.A.D. & Rangel-Negrín, A. 2013. Diets of howler monkeys. In: *Howler Monkeys: Examining the Evolution, Physiology, Behavior, Ecology and Conservation of the Most Widely Distributed Neotropical Primate*, M. Kowalewski, P.A. Garber, L. Cortés-Ortiz, B. Urbani & D. Youlatos (eds.), pp.21–56. Springer, New York.

Di Fiore, A., Link, A. & Dew, J.L. 2008. Diets of wild spider monkeys. In: *Spider monkeys: Behavior, Ecology and Evolution of the Genus* Ateles, C.J. Campbell (ed.), pp.81–137. Cambridge University Press, New York.

dos Reis, N.R., Peracchi, A.L. Batista, C.B. & Rosa, G.L.M. 2015. *Primatas do Brasil: Guia de Campo*. Technical Books Editora, Rio de Janeiro.

Ducke, A. & Black, G. 1953. Phytogeographical notes on the Brazilian Amazon. *An. Acad. Brasil. Ciênc.* **25**: 1–46.

Elliott, A. & Martínez-Vilalta, A. (eds.) 2020. *Mammals of South Asia. Afghanistan, Pakistan, India, Nepal, Bhutan, Bangladesh, Sri Lanka*. Lynx Edicions, Barcelona

Emmons, L.H. 1997. *Neotropical Rainforest Mammals: A Field Guide*. 2nd edition. University of Chicago Press, Chicago, IL.

Fearnside, P.M. 2005. Deforestation in Brazilian Amazonia; history, rates, and consequences. *Conserv. Biol.* **19(3)**: 680–688.

Fernandez-Duque, E. (ed.) 2023. *Owl Monkeys: Biology, Adaptive Radiation, and Behavioral Ecology of the Only Nocturnal Primate in the Americas*. Springer, New York.

Fleagle, J.G. 2000. The century of the past: one hundred years in the study of primate evolution. *Evol. Anthropol.* **9**: 87–100.

Fleagle, J.G., Kay, R.F. & Anthony, M.R.L. 1997. Fossil New World monkeys. In: *Vertebrate Paleontology in the Neotropics*, R.F. Kay, R.H. Madden, R.L. Cifelli and J.L. Flynn (eds.), pp.473–495. Smithsonian Institution Press, Washington, DC.

Ford, S.M., Porter, L.M. & Davis, L.C. (eds.) 2009. *The Smallest Anthropoids: The Marmoset/Callimico Radiation*. Springer, New York.

Fragaszy, D., Fedigan, L. & Visalberghi, E. 2004. *The Complete Capuchin: The Biology of the Genus* Cebus. Cambridge University Press, Cambridge, UK.

Francis, C.M. 2008. *A Guide to the Mammals of Southeast Asia*. Princeton University Press, Princeton, NJ.

Galindo Leal, C. & Câmara, I. de G. 2003. *The Atlantic Forest of South America: Biodiversity Status, Threats, and Outlook*. Island Press and Conservation International, Washington, DC.

Garber, P.A. & Rehg, J.A. 1999. The ecological role of the prehensile tail in white-faced capuchins (*Cebus capucinus*). *Am. J. Phys. Anthropol.* **110**: 325–339.

Gardner, A.L. (ed.) 2007. *Mammals of South America, Volume 1. Marsupials, Xenarthrans, Shrews and Bats*. University of Chicago Press, Chicago, IL.

Gautier-Hion, A., Colyn, M. & Gautier, J.-P. 1999. *Histoire naturelle des Primates d'Afrique Centrale*. Ecofac, Libreville, Gabon.

Goulding, M., Barthem. R. & Ferreira, E. 2003. *The Smithsonian Atlas of the Amazon*. Smithsonian Institution, Washington, DC.

Gray, J. E. 1866. Notice of some new species of spider monkeys (*Ateles*) in the collection of the British Museum. *Proc. Zool. Soc. Lond.* **(1865)**: 732–733.

Groves, C.P., Nguyen Vinh Thanh & Dong Thanh Hai. 2020. *Mammals of Vietnam. Volume 1*. Publishing House for Science and Technology, Vietnam Academy of Science and Technology, Hanoi, Vietnam.

Gusmão, A.C., Messias, M.R. de, Carneiro, J.C., Schneider, H., Alencar, T.B. de, Calouro, A.M., Dalponte, J.C., Mattos, F. de S., Ferrari, S.F., Buss, G., Azevedo, R.B. de, Santos Jr, E. M., Nash, S.D., Rylands A.B. & Barnett, A.A. 2019. A new species of titi monkey, *Plecturocebus* Byrne *et al.*, 2016 (Primates, Pitheciidae) from southwestern Amazonian, Brazil. *Primate Conserv.* **(33)**: 21–35.

Halenar, L.B. & Rosenberger, A.L. 2013. A closer look at the "*Protopihecus*" fossil assemblages: new genus and species from Bahia, Brazil. *J. Hum Evol.* **65**: 374-390.

Harcourt, C.S. & Sayer, J.A. (eds.) 1996. *The Conservation Atlas of Tropical Forests. The Americas*. Simon & Schuster, New York.

Hernández-Camacho, J. & Cooper, R.W. 1976. The non-human primates of Colombia. In: *Neotropical Primates: Field Studies and Conservation*, R.W. Thorington Jr. & P. G. Heltne (eds.), pp.35–69. National Academy of Sciences, Washington, DC.

Hershkovitz, P. 1966. Taxonomic notes on tamarins, genus *Saguinus* (Callithricidae, Primates) with descriptions of four new forms. *Folia Primatol.* **4**: 381–395.

Hershkovitz, P. 1970. Notes on Tertiary platyrrhine monkeys and description of a new genus from the late Miocene of Colombia. *Folia Primatol.* **12**: 1–37.

Hershkovitz, P. 1977. *Living New World Monkeys (Platyrrhini) with an Introduction to Primates, Vol. 1*. Chicago University Press, Chicago.

Jardim, M.M.A., Queirolo, D., Peters, F.B., Mazim, D., Favarini, M.O., Tirelli, F.P., Trindade, R.A., Bonatto, S.L., Bicca-Marques, J.C. & Mourthe, I. 2020. Southern extension of the geographic range of black and gold howler monkeys (*Alouatta caraya*). *Mammalia* **84(1)**: 102–106.

Kay, R.F. 2015. Biogeography in deep time – What do phylogenetics, geology, and paleoclimate tell us about early platyrrhine evolution? *Mol. Phylogenet. Evol.* **82(B)**: 358–374.

Kay, R.F., Meldrum, D.J. & Takai, M. 2013. Pitheciidae and other platyrrhine seed predators. In: *Evolutionary Biology and Conservation of Titis, Sakis and Uacaris*, L. Veiga, A.A. Barnett, S.F. Ferrari and M.A. Norconk (eds.), pp.3–12. Cambridge University Press, Cambridge, UK.

Kellogg, R. & Goldman, E.A. 1944. Review of the spider monkeys. *Proc. U. S. Natl. Mus.* **96**: 1–45.

Kingdon, J. 2015. *The Kingdon Field Guide to African Mammals*. Bloomsbury, London.

Kowalewski, M.M., Garber, P.A., Cortés-Ortiz, L., Urbani, B. & Youlatos, D. (eds.), *Howler Monkeys. Volume 1. Adaptive Radiation, Systematics, and Morphology. Volume 2. Behavior, Ecology, and Conservation*. Springer, New York.

Kuderna, L.F.K. et al. 2023. A global catalogue of whole-genome diversity from 233 primate species. *Science* **380**: 906–913.

Linares, O.J. 1998. *Mamíferos de Venezuela*. Sociedade Conservacionista Audubon de Venezuela, Caracas.

Link, A. & Di Fiore, A. 2006. Seed dispersal by spider monkeys and its importance in the maintenance of Neotropical rain-forest diversity. *J. Trop. Ecol.* **22(3)**: 235–246.

Lynch, J.W., Silva Jr, J.S. & Rylands, A.B. 2012. How different are robust and gracile capuchin monkeys? An argument for the use of *Sapajus* and *Cebus*. *Am. J. Primatol.* **74**: 273–286.

Macfie, E.J. & Williamson, E.A. 2010. *Best Practice Guidelines for Great Ape Tourism*. IUCN SSC Primate Specialist Group, Gland, Switzerland.

Maldonado, A.M., Soto-Calderón, I.D., Hinek, K., Moreno-Sierra, A.M., Lafon, T., Londoño, D., Peralta-Aguilar, A., Inga-Díaz, G., Sánchez, N. & Mendoza, P. 2023. Conservation status of the Nancy Ma's owl monkey (*Aotus nancymaae* Hershkovitz, 1983) on the Colombian-Peruvian border. In: *Owl Monkeys: Biology, Adaptive Radiation, and Behavioral Ecology of the Only Nocturnal Primates in the Americas*, E. Fernandez-Duque (ed.), pp.623–647. Springer, New York.

Marineros, L. & Gallegos, F.M. 1998. *Guía de Campo de los Mamíferos de Honduras*. INADES, Tegucigalpa, Honduras.

Marivaux, L., Negri, F.R., Antoine, P.O., Stutz, N.S., Condamine, F.L., Kerber, L., Pujos, F., Ventura Santos, R., Alvim, A. M. V., Hsiou, A. S., Bissaro Jr, M. C., Adami-Rodrigues, K. & Ribeiro, A.M. (2023). An eosimiid primate of South Asian affinities in the Paleogene of Western Amazonia and the origin of New World monkeys. *Proc. Natl. Acad. Sci. U.S.A*. **120(28)**: e2301338120.

Marsh, L.K. 2014. A taxonomic revision of the saki monkeys, *Pithecia* Desmarest, 1804. *Neotrop. Primates* **21(1)**: 1–163.

Méndez-Carvajal, P.G. 2021. El mono araña negro del Darién es encontrado después de 70 años. *Imagina* **15**: 51–52.

Méndez-Carvajal, P. & Cortés-Ortiz, L. 2020. *Ateles geoffroyi* ssp. *grisescens*. The IUCN Red List of Threatened Species 2020: e.T2287A17979753.

McPhee, R.D.E. & Horovitz, I. 2002. Extinct Quaternary platyrrhines of the Greater Antilles and Brazil. In: *The Primate Fossil Record*, W.C. Hartwig (ed.), pp.189–200. Cambridge University Press, Cambridge, UK.

Mittermeier, R.A. 1977. *A Global Strategy for Primate Conservation*. IUCN SSC Primate Specialist Group, Cambridge, MA. 325 pp.

Mittermeier, R. A. & Konstant, W. R. 2000. Top 25 Most Endangered Primates Launch: Report. Conservation International and IUCN/SSC Primate Specialist Group, Washington, DC.

Mittermeier, R.A. Tattersall, I., Konstant, W.R., Meyers. D.M. & Mast, R.B. 1994. *Lemurs of Madagascar*. Conservation International, Washington, DC.

Mittermeier, R.A., Konstant, W.R., Hawkins, F., Louis, E.E., Langrand, O., Ratsimbazafy, J., Rasoloarison, R., Ganzhorn, J.U., Rajaobelina, S., Tattersall. I. & Meyers, D. 2006. *Lemurs of Madagascar*. 2nd edition. Conservation International, Washington, DC.

Mittermeier, R.A., Louis Jr, E.E., Richardson, M., Schwitzer, C., Langrand, O., Rylands, A.B., Hawkins, F., Rajaobelina, S., Ratsimbazafy, J., Rasoloarison, R., Roos, C., Kappeler, P.M. & MacKinnon, J. 2010. *Lemurs of Madagascar*. 3rd edition, Conservation International, Arlington, VA.

Mittermeier, R.A., Louis Jr, E.E., Langrand, O., Schwitzer, C., Gauthier, C.A., Rylands, A.B., Rajaobelina, S., Ratsimbazafy, J., Rasoloarison, R., Hawkins, F., Roos, C., Richardson, M. & Kappeler, P.M. 2014. *Lémuriens de Madagascar*. 4th edition. Muséum national d'Histoire naturelle, Paris, and Conservation International, Arlington, VA.

Mittermeier, R.A., Reuter, K.E., Rylands, A.B., Louis Jr, E.E., Ratsimbazafy, J., Rene de Roland, L.-A., Langrand, O., Schwitzer, C., Johnson, S.E., Godfrey, L.R., Blanco, M.B., Borgerson, C., Eppley, T.M., Andriamanana, T., Volampeno, S., Adnriantsaralaza, S., Wright, P.C. & Rajaobelina, S. 2023. *Lemurs of Madagascar*. 5th edition. Re:wild, Austin, TX and Lynx Edicions, Barcelona.

Moraes, A.M., Vancine, M.H., Moraes, A.M., Cordeiro, C.L. de O., Pinto, M.P., Lima, A.A., Culot, L., Silva, T.S.F., Collevatti, R.G., Ribeiro, M.C. & Sobral-Souza, T. 2019. Predicting the potential hybridization zones between native and invasive marmosets within Neotropical biodiversity hotspots. *Global Ecol. Conserv.* **20**: e00706.

Napier, P.H. 1976. *Catalogue of the Primates in the British Museum (Natural History). Part I. Families Callitrichidae and Cebidae*. British Museum (Natural History), London.

Norconk, M.A. 2020. Historical antecedents and recent innovations in pitheciid (titi, saki, and uakari) feeding ecology. *Am. J. Primatol.* 83: e23177.

Norconk, M.A. & Setz, Z.S. 2013. Ecology and behavior of saki monkeys (genus *Pithecia*). In: *Evolutionary Biology and Conservation of Titis, Sakis and Uacaris*, L.M. Veiga, A.A. Barnett, S.F. Ferrari & M.A. Norconk (eds.), pp.262–271. Cambridge University Press, Cambridge, UK.

Oates, J.F. 2011. *Primates of West Africa: A Field Guide and Natural History*. Tropical Field Guide Series, Conservation International, Washington, DC.

Ocampo-Peñuela, N. & Winton, S. 2017. Economic and conservation potential bird-watching tourism in post-conflict Colombia. *Trop. Conserv. Sci.* **10(1)**: 1–6.

Oliveira, L.C. & Grelle, C.E.V. 2012. Introduced primate species of an Atlantic Forest region in Brazil: present and future implications for the native fauna. *Trop. Conserv. Sci.* **5(1)**: 112-120.

Organ, J.M., Muchlinski, M.N. & Deane, A.S. 2011. Mechanoreceptivity of prehensile tail skin varies between ateline and cebine primates. *Anat. Rec.* **294**: 2064–2072.

Patton, J.L., Pardiñas, U.F.J. & D'Elía, G. 2015. *Mammals of South America. Volume 2. Rodents.* University of Chicago Press, Chicago, IL.

Peres, C.A., Patton, J.L. & da Silva, M.N.F. 1996. Riverine barriers and gene flow in Amazonian saddle-back tamarins. *Folia Primatol.* **67(3)**: 113–124.

Phillipps, Q. & Phillipps, K. 2018. *Phillipps' Field Guide to the Mammals of Borneo and Their Ecology. Sabah, Sarawak, Brunei and Kalimantan*. 2nd edition. Jean Beaufoy, Oxford.

Reid, F. A. 1967. *A Field Guide to the Mammals of Central America and Southeast Mexico*. Oxford University Press, New York.

Reuter, K.E., Mittermeier, R.A., Williamson, E.A., Jerusalinsky, L., Refisch, J., Sunderland-Groves, J., Byler, D., Konstant, W.R., Vercillo, U.E., Schwitzer, C. & Rylands, A.B. 2022a. Impact and lessons learned from a half-century of primate conservation action planning. *Diversity* **14**: 751. 26pp. DOI 10.3390/d14090751.

Reuter, K.E., Mittermeier, R.A., Schwitzer, C., McCabe, G.M., Rylands, A.B., Jerusalinsky, L., Konstant, W.R., Kerhoas, D., Ratsimbazafy, J., Strier, K.B., Webber, A.D., Williamson, E.A. & Wise, J. 2022b. The 25 most endangered primates list: impacts on conservation fundraising and policy. In: *Communicating Endangered Species: Extinction, News and Public Policy*, E. Freedman, S. S. Hiles & D. Sachsman (eds.), pp.101–115. Routledge, Abingdon, UK.

Rizzini, C.T., Coimbra-Filho, A.F. & Houaiss, A. 1988. *Ecossistemas Brasileiros / Brazilian Ecosystems*. Editora Index, Rio de Janeiro.

Rosenberger, A.L. 2020. *New World Monkeys: The Evolutionary Odyssey.* Princeton University Press, Princeton, NJ.

Rosenberger, A.L. & Tejedor, M.F. 2023. Why owl monkeys are pitheciids: morphology, adaptations and the evolutionary history of the *Aotus* lineage. In: *Owl Monkeys: Biology, Adaptive Radiation, and Behavioral Ecology of the Only Nocturnal Primate in the Americas*, Fernandez-Duque, E. (ed.), pp.103–154. Springer, New York.

Ruschi, A. 1964. Macacos do estado do Espírito Santo. *Bol. Mus. Biol. Prof. Mello Leitao Zool.* **23A**: 1–23.

Rylands, A. B. & Mittermeier, R. A. 2024. Taxonomic database of the IUCN SSC Primate Specialist Group (PSG). Primate Program, Re:wild, Austin, TX. June 2024.

Rylands, A.B., Heymann, E.W., Lynch-Alfaro, J.W., Buckner, J., Roos, C., Matauschek, C., Boubli, J.P., Sampaio R. & Mittermeier, R.A. 2016. Taxonomic review of the New World tamarins (Callitrichidae, Primates). *Zool. J. Linn. Soc.* **177**: 1003–1028.

Schneider, H. & Rosenberger, A.L. 1996. Molecules, morphology, and platyrrhine systematics. In: *Adaptive Radiations of Neotropical Primates*, M.A. Norconk, A.L. Rosenberger and P. A. Garber (eds.), pp.3–19. Springer, New York.

Schrago, C.G. & Russo, C.A.M. 2003. Timing the origin of New World Monkeys. *Mol. Biol. Evol.* **20(10)**: 1620–1625.

Schrago, C.G., Menezes, A.N., Furtado, C., Bonvicino, C.R. & Seuánez, H.N. 2014. Multispecies coalescent analysis of the early diversification of Neotropical primates: phylogenetic inference under strong gene trees/species tree conflict. *Genome Biol. Evol.* **6**: 3105–3114.

Schwoerer, T. & Dawson, N.G. 2022. Small sight – big might: economic impact of bird tourism shows opportunities for rural communities and biodiversity conservation. *PLoS One* **17(7)**: e0268594.

Sclater, P.L. 1858. On the general geographic distribution of the members of the Class Aves. *Zool. J. Linn. Soc.* **2(7)**: 130–145.

Seiffert, E.R., Tejedor, M.F., Fleagle, J.G., Novo, N.M., Cornejo, F.M., Bond, M., de Vries, D. & Campbell Jr, K.E. 2020. A parapithecid stem anthropoid of African origin in the Paleogene of South America. *Science* **368**: 194–197.

Shepherd, C.R. & Shepherd, L.A. 2017. *A Naturalist's Guide to Primates of Southeast Asia, East Asia and the Indian Subcontinent*. John Beaufoy Publishing, Oxford, UK.

Silvestro, D., Tejedor, M.F., Serrano, M.L., Loiseau, O., Rossier, V., Rolland, J., Zizka, A., Höhna, S., Antonelli, A. & Salamin, N. 2019. Early arrival and climatically-linked geographic expansion of New World Monkeys from tiny African ancestors *Syst. Biol.* **68(1)**:78–92.

Smith, A. & Yan Xie (eds.) 2008. *A Guide to the Mammals of China*. Princeton University Press, Princeton, NJ.

Smith, P., Rios, S.D. & Smith, R.L. 2021. Paraguayan primatology: past, present and future. *Primate Conserv.* **(35)**: 47–68.

Starin, D. 2009. Please don't feed the monkeys: tourism alters primate behavior in a Gambian forest park. *The Wildlife Professional*. Summer 2009, pp.54–57. The Wildlife Society.

Strier, K.B. 1999. *Faces in the Forest*. Harvard University Press, Cambridge, MA.

Strier, K.B., Melo, F.R., Mendes, S.L., Valença-Montegro, M.M., Rylands, A.B., Mittermeier, R.A. & Jerusalinsky, L. 2021. Science, policy, and conservation management for a Critically Endangered primate in the Atlantic Forest of Brazil. *Front. Cons. Sci.* **2**: 734183.

Supriatna, J. 2019. *The Field Guide to the Indonesia Primates*. Yayasan Pustaka Obor, Indonesia.

Tabarelli, M., Pinto, L.P., Silva, J.M.C., Hirota, M. & Bedê, L. 2005. Challenges and opportunities for biodiversity conservation in the Brazilian Atlantic Forest. *Conserv. Biol.* **19(3)**: 695–700.

Tejedor, M.F. & Novo N.M. 2017. Origen e historia evolutiva de los primates platirrinos: nuevas evidencias. In: *La Primatología en Latinoamérica 2. Tomo I. Argentina-Colombia*, B. Urbani, M. Kowalewski, R.G.T. da Cunha, S. de la Torre and L. Cortés-Ortiz (eds.), pp.27–38. Ediciones IVIC, Instituto Venezolano de Investigaciones Científicas, Caracas, Venezuela.

Tirira, D. 2007. *Mamíferos del Ecuador: Guía de Campo*. Ediciones Murciélago Blanco, Quito.

Tomas, W.M., Timo, T.P.C., Camilo, A.R., Oliveira, M.R., Tortato, F.R., Mamede, S., Benites, M., Garcia, C.M., Gusmão, A.C. & Rimoli, J. 2022. Primatas ocorrentes na Bacia do Alto Paraguai e Pantanal, Brasil. *Bol. Mus. Para. Emílio Goeldi. Cienc. Nat.*, Belém **17(3)**: 701–724.

Veiga, L.M. & Ferrari, S.F. 2013. Ecology and behavior of bearded sakis (genus *Chiropotes*). In: *Evolutionary Biology and Conservation of Titis, Sakis and Uacaris*, L.M. Veiga, A.A. Barnett, S.F. Ferrari & M.A. Norconk (eds.), pp.240–249. Cambridge University Press, Cambridge, UK.

Veiga, L.M., Barnett, A.A., Ferrari, S.F. & Norconk, M.A. (eds.) 2013. *Evolutionary Biology and Conservation of Titis, Sakis and Uacaris*. Cambridge University Press, Cambridge, UK.

Walker, S. & Molur, S. 2007. *Guide to South Asia Primates for Teachers and Students of all Ages*. Zoo Outreach Organisation, PSG South Asia and WILD, Coimbatore, India.

Wallace, A.R. 1876. *The Geographical Distribution of Animals with a Study of the Relations of Living and Extinct Faunas as Elucidating the Past Changes of the Earth's Surface*. 2 volumes. Macmillan and Co., London.

Waters, S., Setchell, J.M., Maréchal, L., Oram, F., Wallis, J. & Cheyne, S.M. 2021. *Best Practice Guidelines for Responsible Images of Non-human Primates*. IUCN SSC Primate Specialist Group Section on Human-Primate Interactions. Available online: human-primate-interactions.org/resources.

Waters, S., Hansen, M.F., Setchell, J.M., Cheyne, S.M., Mittermeier, R. A., Ang, A., Aldrich, B.C., Andriantsaralaza, S., Clarke, T.A., Dempsey, A., Dore, K. M., Hanson, K. T., Kitegile, A., Maldonado, A. M., Maréchal, L., McKinney, T., Ruiz Miranda, C. R., Niu, K., Svensson, M.S., Talebi, M., Wallis, J., Williams, J., Gursky, S., Peng-Fei, F., Chetry, D. & Behie, A. 2023a. *Responsible Primate-watching for Tourists*. IUCN SSC Primate Specialist Group Section on Human-Primate Interactions. Available online: human-primate-interactions.org/resources.

Waters, S., Hansen, M.F., Setchell, J.M., Cheyne, S.M., Ampumuza, C., Hanson, K.T., Hockings, K.J., Ilham, K., Kaburu, S.S.K., Maréchal, L., Ruiz Miranda, C.R. & Wallis, J. 2023b. *Responsible Primate-watching for Primate Tourism Professionals*. IUCN SSC Primate Specialist Group Section on Human-Primate Interactions. Available online: human-primate-interactions.org/resources.

Whitmore, T.C. & Prance, G.T. (eds.) 1986. *Biogeography and Quaternary History in Tropical America*. Clarendon Press, Oxford.

Withrow, B. 2019. Millions of birders around the world set out to catch a rare glimpse of plumage, a bold stroke of color, or to hear an unusual song – and they are changing the face of tourism. *Daily Beast*, July 23, 2019.

USING THE ILLUSTRATED CHECKLIST

Background

The compilation of the nine volumes of the *Handbook of the Mammals of the World* (HMW) series saw the first volume published in 2009 and the last in 2019. A major global effort, this series of books involved gathering a wealth of information on all the world's mammal species in little over a decade. In 2020, the series was summarized in a comprehensive and practical two-volume set entitled *Illustrated Checklist of the Mammals of the World*. Subsequently, in 2023, Lynx produced a single volume entitled *All the Mammals of the World*, which provides an illustrated overview of mammalian diversity.

In 2020, using the information compiled for the *Handbook* and the *Illustrated Checklist*, Lynx launched a series of regional illustrated checklists focused on important geographical areas to help naturalist-travellers assess with ease how many mammal species are present in a given area, help them identify what they see, and enable them to create or complement their own mammal life-lists. These regional guides provide brief descriptions including size and weight, habitats used by each species, maps of their geographical ranges, and a colour illustration of each species.

Four regional illustrated checklists have already appeared in this series, *Mammals of the Southern Cone—Argentina, Chile, Paraguay and Uruguay* (February 2020), *Mammals of South Asia—Afghanistan, Pakistan, India, Nepal, Bhutan, Bangladesh, and Sri Lanka* (June 2020), *Mammals of China* (October 2020), and *Mammals of Madagascar* (May 2021). This last was produced in collaboration with Re:wild and here we have the fifth guide in this series, *Neotropical Primates*, also produced in collaboration with Re:wild.

Geographical scope

The geographical scope of this guide is the Neotropical Region, one of the six major biographical regions of the world as defined by its characteristic fauna and flora. Primates occur in a large part of this region today (in all but Chile of the continental countries) that extends from southern Mexico through Central America to northern Argentina and southern Brazil, and just very marginally into northern Uruguay, as well as on a few islands, for example, Trinidad, Margarita (Venezuela), Gorgona (Colombia) and Coiba (Panama).

Associated with the slave trade, three African primates have been introduced onto the islands of Barbados, St. Kitts and Nevis, and St. Lucia (Green Monkey, *Chlorocebus sabaeus*), Grenada (Mona Monkey, *Cercopithecus mona*), and Saint-Martin/Sint Maarten and Anguilla (Vervet Monkey, *Chlorocebus pygerythrus*) (Beck, 2018; Jones *et al.*, 2018). There was also once a small number of vervets on the island of Sint Eustatius (also called Statia) but they have been removed. In Africa, the Green Monkey is native to Senegal, Gambia, Guinea-Bissau, Guinea and Sierra Leone east to the Nakambe/Volta River in Burkina Faso and Ghana (Oates, 2011). The six subspecies of the Vervet Monkey are widespread in East and southern Africa from Ethiopia to South Africa (De Jong & Butynski, 2023). The Mona Monkey is also West African and is found in forests just west of the Volta River in Ghana, and east through Toga, Benin, southern Nigeria and northern Cameroon (Oates, 2011).

Taxonomic treatment

The systematics and taxonomy used in this illustrated checklist largely follow the *Handbook of the Mammals of the World* (HMW) (Mittermeier *et al.*, 2013) and the 2020 two-volume *Illustrated Checklist of the Mammals of the World* (Burgin *et al.*, 2020) but have been updated following the database of the IUCN SSC Primate Specialist Group.

Changes at the genus level

Hershkovitz (1977) divided the tamarins, genus *Saguinus* Hoffmannsegg, into six species groups: the white-mouthed or *nigricollis* group, the moustached or *mystax* group, the Colombian and Panamanian bare-faced tamarins or *oedipus* group, the Golden-handed Tamarin or *midas* group, the Mottled-face tamarin group (*inustus*), and the Brazilian bare-faced tamarins or *bicolor* group. Following the phylogenetic study of the callitrichids by Buckner *et al.* (2015), Rylands *et al.* (2016) placed the white-mouthed tamarins in the genus *Leontocebus* Wagner. Brcko *et al.* (2022) went further and placed the moustached tamarins in the genus *Tamarinus* Trouessart, and the Colombian and Panamanian bare-faced tamarins in the genus *Oedipomidas* Reichenbach. The Golden-handed tamarin and Brazilian bare-faced tamarin groups now make up the genus *Saguinus*. Despite lacking a moustache, the Mottled-face Tamarin, *inustus*, has been found to be a member of the moustached tamarin clade (see Buckner *et al.*, 2015) and so is now placed with them in the genus *Tamarinus*.

New species

Thirteen new species have been described since 2014 (Rylands & Mittermeier, 2024b). Three are in the Family Callitrichidae: the Munduruku Marmoset, *Mico munduruku* Costa-Araújo *et al.*, 2019, between the rios Tapajós and Jamanxim in the south of the Brazilian state of Para; Schneider's Marmoset, *Mico schneideri* Costa-Araújo *et al.*, 2021, between the rios Juruena and Teles-Pires in the south-east of the Amazon basin in Brazil; and Kulina's Moustached Tamarin, *Saguinus kulina* Lopes *et al.*, 2023 (now in the genus *Tamarinus*), from between the lower Rio Juruá and the Rio Tefé, Brazil.

Four of the new species are titi monkeys, members of the Family Pitheciidae: Milton's Titi, *Plecturocebus miltoni* (Dalponte *et al.*, 2014), from the left bank of the Rio Aripuanã in the Brazilian state of Mato Grosso; the Urubamba Brown Titi, *Plecturocebus urubambensis* (Vermeer & Tello-Alvarado, 2015), from the Río Manu in south-eastern Peru; Groves's Titi *Plecturocebus grovesi* Boubli *et al.*, 2018, between the Rio Teles-Pires and the rios Juruena and Arinos in the north of the state of Mato Grosso, Brazil; and Aquino's Collared Titi, *Cheracebus aquinoi* Rengifo *et al.*, 2023, from between the ríos Nanay and Tigre, right bank affluents of the lower Río Ucayali, in Peru. Vermeer & Tello-Alvarado (2015) also resurrected the name *Callicebus toppini* Thomas, 1914 (now of the genus *Plecturocebus*) as the correct name for the titi monkey occurring between the right (south) bank of the middle-to-upper Rio Purus, west from the Rio Ituxi, west to the Río Manu and south to the ríos Madre de Dios and Tambopata, formerly thought to have been occupied by the Doubtful Titi, *Plecturocebus dubius* (see Van Roosmalen *et al.*, 2005).

Five saki monkeys, genus *Pithecia*, also of the Family Pitheciidae, were described as distinct species in a major taxonomic revision by Marsh published in 2014: Isabel's Saki *P. isabela* in the Pacaya-Samiria National Reserve between the lower ríos Ucayali and Marañón in northern Peru; Cazuza's Saki, *P. cazuzai*, between the lower Rio Japurá and Rio Solimões in Brazil; Mittermeier's Tapajós Saki, *P. mittermeieri*, between the rios Madeira and Tapajós; Rylands's Bald-faced Saki, *P. rylandsi*, in south-eastern Peru, northern Bolivia and the Brazilian state of Rondônia; and Pissinatti's Bald-faced Saki, *P. pissinattii*, between the lower rios Madeira and Purus. Serrano-Villavicencio *et al.* (2019) reviewed the evidence for these three final species—*mittermeieri*, *rylandsi* and *pissinattii*—and concluded that they are junior synonyms of *irrorata*. Genetic research currently underway may confirm their conclusion (J.P. Boubli, pers. comm.). The distribution of *P. rylandsi* indicated by Marsh (2014) evidently overlaps with *P. irrorata*, a very unlikely situation, so for this reason alone here we consider *P. rylandsi* to be a junior synonym.

Finally, and again in the Pitheciidae, is the Kanamari Bald Uacari, *Cacajao amuna* Silva *et al.*, 2022, from the Rio Tarauacá, a right-bank tributary of the middle Rio Juruá in Brazil.

Taxonomic re-arrangements and nomenclatural changes

In addition to these new species, there have been a number of taxonomic re-arrangements and nomenclatural changes for certain members of the Family Callitrichidae (marmosets and tamarins) and the bald uacaris. Formerly, Spix's Moustached Tamarin was believed to comprise three subspecies, the nominate form *Saguinus mystax mystax*, *S. m. pileatus* and *S. m. pluto*. A genetic study by Lopes *et al.* (2023a), however, showed that *S. mystax* is a monotypic species and that *pluto* is correctly a subspecies of *S. pileatus*. These tamarins are now placed in the genus *Tamarinus*. Gregorin *et al.* (2023) determined that the Bearded Emperor Tamarin, formerly *Saguinus imperator subgrisescens*, is in fact a distinct species. These tamarins are now placed in the genus *Tamarinus*. In a phylogenetic study of all the tamarins, Lopes *et al.* (2023b) found that Thomas's Red-bellied Tamarin, considered to be a subspecies of *Tamarinus labiatus*, is in fact a distinct species, *T. thomasi*. Lopes *et al.* (2023b) also confirmed that the two emperor tamarins are good species and that *Leontocebus fuscus*, considered by Hershkovitz (1977) to be a subspecies of *Saguinus* (now *Leontocebus*) *fuscicollis*, is a sister species to *L. nigricollis*.

In 2011, Matauschek *et al.* carried out a phylogenetic analysis of the Peruvian tamarins and concluded that all but one of the subspecies of Spix's Saddle-back Tamarin, *Saguinus* [now *Leontocebus*] *fuscicollis*, should be considered species. The exception was the White Saddle-back Tamarin *melanoleucus*, which remained as a subspecies, not of *fuscicollis* but of *weddelli*, which the authors had found to be a very close relative. The Brazilian endemic subspecies—*avilapiresi*, *cruzlimai*, *primitivus* and *mura*—were not included in this analysis (2011). Sampaio *et al.* (2015) subsequently showed *cruzlimai* to be a distinct species, and here we list the remaining three Brazilian forms, *avilapiresi*, *primitivus* and *mura*, as full species. This change has already been signalled by Fabio Röhe in his doctoral thesis (2021) on the systematics and biogeography of the genus *Leontocebus*. Röhe also placed *L. melanoleucus* as a distinct species and this is currently being confirmed by further genetic studies (A.B. Martins, pers. comm.).

An analysis of the mitochondrial DNA of the genus *Cebus* by Boubli *et al.* (2012) resulted in the finding that the Venezuelan capuchin, hitherto known as *Cebus brunneus*, was a distinct species with affinities to the white-fronted capuchins (the group that includes *Cebus albifrons*) rather than to the weeper or wedge-capped capuchins (the group that includes *Cebus olivaceus*) as had been thought previously. It was listed in Mittermeier *et al.* (2013). Examination of the type specimen (the first specimen to which the name was attributed) revealed that it was in fact *C. olivaceus*, thus rendering the name *brunneus* invalid. Boubli *et al.* (2012) were not wrong, however, in identifying a distinct capuchin, albeit lacking a name, in the white-fronted group in Venezuela, and studies are now underway to describe it, determine its geographical distribution, and give it a name (B. Urbani, J.W. Lynch, pers. comm).

In recent years a number of phylogenetic studies of the robust capuchins have shown that the many subspecies that have been described for the Amazonian Brown Capuchin, *Sapajus apella*, are in fact too closely related to be considered valid. For a time, the Large-headed Capuchin of western Amazonia was tentatively considered to be a subspecies but further genetic analyses by Martins (2021) have shown it to be a distinct species, *Sapajus macrocephalus*.

Vermeer & Tello-Alvarado (2015) resurrected the name *Callicebus toppini* Thomas, 1914 (now of the genus *Plecturocebus*) as the correct name for the titi monkey occurring between the right (south) bank of the middle to upper Rio Purus, west from the Rio Ituxi, west to the Río Manu, and south to the ríos Madre de Dios and Tambopata in Peru and Bolivia (Martínez *et al.*, 2024), formerly thought to have been occupied by the Doubtful Titi, *Plecturocebus dubius* (see Van Roosmalen *et al.*, 2005).

The White-collared Titi *Cheracebus torquatus*, named by Hoffmannsegg in 1807, was the first widow monkey to be described. Since then, however, it has had a complex taxonomic history, its distribution was never clearly defined due to the lack of a type locality, and nobody knew from where it had been collected. Hershkovitz (1990) suggested that it occurred along the north bank of the Rio Solimões despite there being no evident delineation with the range of the distinct *C. lugens* that is known to occur there. A broad and detailed analysis of museum specimens by Byrne *et al.* (2020) finally resolved its distribution as between the rios Purus and Juruá in Brazil (south, not north, of the Rio Solimões). Thus, *torquatus* is a senior synonym of the name that had been given to the collared titi there, *Callicebus purinus*, by Oldfield Thomas in 1927.

Marsh (2014) revised the taxonomy of the sakis, *Pithecia*, based on an examination of their morphology and pelage patterns and coloration, research that resulted in a list of 16 species. They included the eight taxa recognized by Hershkovitz (1987), two that he considered to be synonyms (*P. hirsuta* and *P. inusta*), one which Hershkovitz had evidently overlooked (*P. napensis*), and the five newly described species (*P. cazuzai*, *P. isabela*, *P. mittermeieri*, *P. rylandsi* and *P. pissinattii*) mentioned above.

Describing the new species of bald uacari, Silva *et al.* (2022) raised the three subspecies of *Cacajao calvus*—*rubicundus*, *ucayalii* and *novaesi*—to species level. Recent genetic studies by Oklander *et al.* (submitted) have shown that the division of the Brown Howler, *Alouatta guariba*, into two subspecies is not valid, and here we consider it to be monotypic, with *Alouatta guariba clamitans* Cabrera, 1940, a junior synonym. We also believe that the Colombian Black Spider Monkey, a distinct and allopatric former subspecies of *Ateles fusciceps*, should be considered a species, *Ateles rufiventris*.

For many years there was indecision regarding the taxonomic status of the northern and southern muriquis, that is, whether they should be classified as species or subspecies. This was resolved in a phylogenetic and phylogeographic analysis by Chaves *et al.* (2019): both forms are valid species, the Southern Muriqui, *Brachyteles arachnoides*, and the Northern Muriqui, *B. hypoxanthus*.

Species accounts

Each account for a species or subspecies begins with its common name in English, its scientific name, and its conservation status according to the IUCN Red List of Threatened Species, as follows:

- **LC** Least Concern
- **NT** Near Threatened
- **VU** Vulnerable
- **EN** Endangered
- **CR** Critically Endangered
- **DD** Data Deficient
- **NE** Not Evaluated

Immediately beneath, in bold, are common names in the two most widely used languages in the Neotropical region, Spanish (ES) and Portuguese (PT), as well as in French (FR) for French Guiana, Sranan Tongo (SR), an English-based creole language, for Suriname and a local language for Guyana (GY). Although different names are available in these languages—and many more are used by native indigenous peoples—we have tried to choose the ones most commonly used or understood.

Each account is accompanied by a distribution map, based on HMW, but updated by the current authors. Distributions are marked in green. Next to the map, a short text indicates in which countries the species or subspecies occurs. The **E** indicates that it is endemic to a particular country. The following two-letter ISO country codes are used: AR (Argentina), BZ (Belize), BO (Bolivia), BR (Brazil), CO (Colombia), CR (Costa Rica), EC (Ecuador), SV (El Salvador), GF (French Guiana), GT (Guatemala), GY (Guyana), HN (Honduras), MX (Mexico), NI (Nicaragua), PA (Panama), PY (Paraguay), PE (Peru), SR (Suriname), TT (Trinidad and Tobago), UY (Uruguay) and VE (Venezuela). The Caribbean islands with introduced African monkeys are AI (Anguilla), BB (Barbados), KN (Saint Kitts and Nevis), LC (Saint Lucia), MF (Saint Martin, French part), SX (Sint Maarten, Dutch part) and GD (Grenada).

The main body of the text starts with the Habitat section, marked with **H**, presenting the most relevant details regarding the habitats used by the species or subspecies, as well as further details about its geographical range. This is followed by the **ID** section, which provides a synthesis of descriptive notes to help with identification. At the end of the text, denoted by a Δ, is the range of altitudes at which the species or subspecies can be found, if available.

Each account includes at least one illustration, with additional illustrations labelled as necessary. Illustrations are not to scale and are reproduced as large as possible to facilitate the appreciation of details. The illustrations are complemented by the most relevant biometric data: **HB** = head-body length, **T** = tail length and **W** = weight. Lengths are given in cm and weights in grams (g) or kilograms (kg) depending on the unit considered to be the most appropriate. Separate information provided for males and females is indicated with (M) for males and (F) for females after the corresponding measurement.

At the end of the book (pp. 126–137), we provide a checklist to be marked with the primates you have seen; the list includes additional common names for each species and subspecies and the countries where they are known to occur, as well as the conservation status.

References

Beck, B. B. 2018. *Unwitting Travellers: A History of Primate Reintroduction.* Saltwater Media, Berlin, MD.
Boubli, J.P., Rylands, A.B., Farias, I., Alfaro, M. & Lynch Alfaro, J. 2012. *Cebus* phylogenetic relationships: a preliminary reassessment of the diversity of the untufted capuchin monkeys. *Am. J. Primatol.* **74**: 381–393.
Boubli, J.P., Byrne, H., Silva, M.N.F. da, Silva-Júnior, J.S, Costa Araujo, R., Bertuol, F., Gonçalves, J., Melo, F.R., Rylands, A.B., Mittermeier, R.A., Silva, F.E., Nash, S.D., Canale, G., Alencar, R. de M., Rossi, R., Carneiro, J., Sampaio, I., Farias, I.P., Schneider, H. & Hrbek, T. 2019. On a new species of titi monkey (Primates: *Plecturocebus* Byrne *et al.*, 2016), from Alta Floresta, southern Amazon, Brazil. *Mol. Phylogenet. Evol.* **132**: 117–137.
Brcko, I.C., Carneiro, J., Ruiz-García, M., Boubli, J.P., Silva-Júnior, J.S., Farias, I., Hrbek, T., Schneider, H. & Sampaio, I. 2022. Phylogenetics and an updated taxonomic status of the tamarins (Callitrichinae, Cebidae). *Mol. Phylogenet. Evol.* **173**: 10754.
Buckner, J.C., Lynch Alfaro, J., Rylands, A.B. & Alfaro, M.E. 2015. Biogeography of the marmosets and tamarins (Callitrichidae). *Mol. Phylogenet. Evol.* **82**: 413–425.
Burgin, C J., Wilson, D.E., Mittermeier, R.A., Rylands, A.B., Lacher, T.E. & Sechrest, W. (eds.) 2020. *Illustrated Checklist of the Mammals of the World.* Two volumes. Lynx Edicions, Barcelona.
Byrne, H., Rylands, A.B., Nash, S.D. & Boubli, J.P. 2020. On the taxonomic history and true identity of the collared titi, *Cheracebus torquatus* (Hoffmannsegg, 1807) (Platyrrhini, Callicebinae). *Primate Conserv.* (**34**):13–52.
Chaves, P.B., Magnus, T., Jerusalinsky, L., Talebi, M., Strier, K.B., Breves, P., Tabacow, F., Teixeira, R.H.F., Moreira, L., Hack, R.O.E., Milagres, A., Pissinatti, A., Melo, F.R. de, Pessutti, C., Mendes, S.L., Margarido, T.C., Fagundes, V., Di Fiore, A. & Bonatto, S.L. 2019. Phylogeographic evidence for two species of muriqui (genus *Brachyteles*). *Am J. Primatol.* **81(12)**: e23066.
Costa-Araújo, R., Melo, F.R. de, Canale, G.R., Hernández-Rangel, S.M., Messias, M.R., Rossi, R.V., Silva, F.E., da Silva, M.N.F., Nash, S.D., Boubli, J.P., Farias, I.P. & Hrbek, T. 2019. The Munduruku marmoset: a new monkey species from southern Amazonia. *PeerJ.* **7**: e7019. DOI 10.7717/peerj.7019. 18pp.
Costa-Araújo, R., Silva-Jr, J.S., Boubli, J.P., Rossi, R.V., Canale, G.R., Melo, F.R., Bertuol, F., Silva, F.E., da Silva, M.N.F., Nash, S.D., Sampaio, I., Farias I.P. & Hrbek, T. 2021. An integrative analysis uncovers a new, pseudo-cryptic species of Amazonian marmoset (Primates: Callitrichidae: *Mico*) from the arc of deforestation. *Sci. Rep.* DOI 10.1038/s41598-021-93943-w

Dalponte, J., Silva, F.E. & Silva-Júnior, J.S. 2014. New species of titi monkey, genus *Callicebus* Thomas, 1903 (Primates, Pitheciidae), from southern Amazonia, Brazil. *Pap. Avuls. Zool., São Paulo* **54(32)**: 457–472.

De Jong, Y. A. & Butynski, T. M. 2023. *Primates of East Africa: Pocket Identification Guide.* Illustrated by Stephen D. Nash. 2nd edition. Tropical Pocket Guide Series, Re:wild, Austin, TX.

Gregorin, R., Athaydes, D., dos Santos-Júnior, J.E. & Ayoub, T.B. 2023. Taxonomic status of *Tamarinus imperator subgrisescens* (Lönnberg, 1940) (Cebidae, Callitrichinae). *Pap. Avuls. Zool., São Paulo* (**63**): e202363005.

Hershkovitz, P. 1977. *Living New World Monkeys (Platyrrhini) with an Introduction to Primates,* Volume 1. Chicago University Press, Chicago.

Hershkovitz, P. 1987. The taxonomy of South American sakis, genus *Pithecia* (Cebidae, Platyrrhini): a preliminary report and critical review with the description of a new species and new subspecies. *Am. J. Primatol.* **12**: 387–468.

Hershkovitz, P. 1990. Titis, New World monkeys of the genus *Callicebus* (Cebidae, Platyrrhini): a preliminary taxonomic review. *Fieldiana, Zool., n.s.* (**55**): 1–109.

Jones, H.P., Campbell, K.J., Burke, A.M., Baxter, G.S., Hanson, C.C. & Mittermeier, R.A. 2018. Introduced non-hominid primates impact biodiversity and livelihoods: management priorities. *Biol. Invasions* 20: 2329–2342.

Lopes, G.P., Röhe, F., Bertuol, F., Polo, E., Lima, I.J., Valsecchi, J., Santos, T.C.M., Nash, S.D., da Silva, M.N.F., Boubli, J.P., Farias, I.P. & Hrbek, T. 2023a. Taxonomic review of *Saguinus mystax* (Spix, 1823) (Primates, Callitrichidae), and description of a new species. *Peer J.* 10.7717/peerj.14726.

Lopes, G.P., Rohe, F., Bertuol, F., Polo, E., Valsecchi, J., Santos, T.C.M., Silva, F.E., Lima, I.J., Sampaio, R., da Silva, M.N.F., Silva, C.R., Boubli, J., Costa-Araújo, R., de Thoisy, B., Ruiz-García, M., Gordo, M., Sampaio, I., Farias, I.P. & Hrbek, T. 2023b. Molecular systematics of tamarins with emphasis on genus *Tamarinus* (Primates, Callitrichidae). *Zool. Scripta* 52: 556–570.

Lynx Nature Books. 2023. *All the Mammals of the World.* Barcelona.

Marsh, L.K. 2014. A taxonomic revision of the saki monkeys, *Pithecia* Desmarest, 1804. *Neotrop. Primates* **21(1)**: 1–163.

Martínez, J., Hernani-Lineros, L. & Rumiz, D.I. 2024. *Los Monos o Primates No Humanas de Bolivia.* Guías de Biodiversidad Boliviana, Fundación Patiño, Santa Cruz de la Sierra, Bolivia.

Martins, A.B. 2021 A Phylogenomics and Population Genomics study of the Robust Capuchin Monkey (*Sapajus*) Radiation: First Genus-wide Analyses of Admixture and Species Boundaries in Neotropical Primates. PhD thesis, University of Texas, Austin, TX.

Matauschek, C., Roos, C. & Heymann, E.W. 2011. Mitochondrial phylogeny of tamarins (*Saguinus*, Hoffmannsegg, 1807) with taxonomic and biogeographic implications for the *S. nigricollis* species group. *Am. J. Phys. Anthropol.* **144**: 564–557.

Mittermeier, R.A., Rylands, A.B. & Wilson, D.E. (eds.) 2013. *Handbook of the Mammals of the World.* Volume 3. Primates. Lynx Edicions, Barcelona. 951pp.

Oates, J.F. 2011. *Primates of West Africa: A Field Guide and Natural History.* Tropical Field Guide Series, Conservation International, Washington, DC.

Oklander, L.I., Fernandez, G.P., Machado, S., Caputo, M., Hirano, Z.M.B., Rylands, A.B., Neves, L.G., Mendes, S.L., Pacca, L.G., de Melo, F.R., Mourthé, I., Freitas, T.R.O., Corach, D., Jerusalinsky, L. and Bonatto, S.L. 2024. Phylogeography, taxonomy and conservation of the endangered brown howler monkey, *Alouatta guariba* (Primates, Atelidae), of the Atlantic Forest. *Front. Genet.* Submitted.

Rengifo, E.M, D'Elia, G., García, G., Charpentier, E. & Cornejo, F.M. 2023. A new species of titi monkey, genus *Cheracebus* Byrne *et al.*, 2016 (Primates: Pitheciidae), from Peruvian Amazonia. *Mammal Study* **48(1)**: 2022-0019.

Röhe, F. 2021. Sistemática Molecular e Biogeografia do Gênero *Leontocebus* Wagner, 1839 (Callitrichidae-Primates). PhD thesis, Instituto Nacional de Pesquisas da Amazônia (INPA), Manaus, AM, Brazil.

Rylands, A.B. & Mittermeier, R.A. 2024. Taxonomy and systematics of the Neotropical primates: a review and update. *Front. Conserv. Sci.* 5: 1391303.

Rylands, A.B., Heymann, E.W., Lynch Alfaro, J.W., Buckner, J., Roos, C., Matauschek, C., Boubli, J.P., Sampaio, R. & Mittermeier, R.A. 2016. Taxonomic review of the New World tamarins (Callitrichidae, Primates). *Zool. J. Linn. Soc.* **177**: 1003–1028.

Sampaio, R., Röhe, F., Pinho, G., Silva-Júnior, J.S., Farias, I.P. & Rylands, A.B. 2015. Re-description and assessment of the taxonomic status of *Saguinus fuscicollis cruzlimai* Hershkovitz, 1966 (Primates, Callitrichinae). *Primates* **56**: 131–144.

Serrano-Villavicencio, J.S., Hurtado, C.M., Vendramel, R.L. & do Nascimento, F.B. 2019. Reconsidering the taxonomy of the *Pithecia irrorata* species group (Primates: Pitheciidae). *J. Mammal.* **100(1)**: 130–141.

Silva, F.E., do Amaral, J.V., Roos, C., Bowler, M., Röhe, F., Sampaio, R., Bertuol, F., Santana, M.I., Silva Júnior, J.S., Rylands, A.B., Hrbek, T. & Boubli, J.P. 2022. Molecular phylogeny and systematics of bald uacaris, genus *Cacajao* Lesson, 1840 (Primates: Pitheciidae), with the description of a new species. *Mol. Phylogenet. Evol.* **173**: 107509.

Van Roosmalen, M.G.M., Van Roosmalen, T. & Mittermeier, R.A. 2005. A taxonomic review of the titi monkeys, genus *Callicebus* Thomas, 1903, with the description of two new species, *Callicebus bernhardi* and *Callicebus stephennashi*, from Brazilian Amazonia. *Neotrop. Primates* **10 (Suppl.)**: 1–52.

Vermeer, J. & Tello-Alvarado, J.C. 2015. The distribution and taxonomy of titi monkeys (*Callicebus*) in central and southern Peru, with the description of a new species. *Primate Conserv.* **(29)**: 9–29.

ACKNOWLEDGEMENTS

We acknowledge the many foundations and people who have supported hundreds of primate conservation projects over the past five decades, which includes a large body of work on Neotropical primates. In particular, we are grateful to the Margot Marsh Biodiversity Foundation, heavily involved in supporting primate conservation since 1996, the Mohamed bin Zayed Species Conservation Fund, which has provided grants since 2009, the World Wildlife Fund–US, which supported a Primate Program from 1979 to 1989, Conservation International, which supported primate projects in 1989–2017, and, most recently, Re:wild (formerly Global Wildlife Conservation), which has maintained an active role in primate research and conservation since 2018.

Thanks are also due to the many generous people who have been dedicated supporters of primate conservation for decades, notably the late Doris and Dale Swanson, Nancy and Dan Jochem, Shawn Concannon, Nick and Isaac Pritzker, John Swift, Mark O'Donnell, Annette Lanjouw and William R. Konstant. Last, and certainly not least, very special thanks to Jon Stryker and Slobodan Randjelović for providing major support for Neotropical primate projects since 2022. Much of the information found in this book is directly attributable to the projects funded over the years by these people and organizations. We are extremely grateful to all of them. Furthermore, Ella Outlaw, Jill Lucena and Stephen Nash provide constant and invaluable support for the endeavours of the IUCN SSC Primate Specialist Group, including membership, grant awards, illustrations, publications and numerous other initiatives and programmes, for which we are eternally grateful.

To close, we also thank Lynx Nature Books and particularly Josep del Hoyo, Amy Chernasky and Albert Martínez-Vilalta, for all that they have done to increase our knowledge of the world's mammals through their amazing nine-volume series *Handbook of the Mammals of the World* (2009–2019), the two volumes of the *Illustrated Checklist of the Mammals of the World* (2020), and *All the Mammals of the World* (2023), as well as the series of regional illustrated checklists, of which this volume is a part. They have made a great contribution to our understanding of and appreciation for this hugely important component of our planet's unique biodiversity.

Species accounts

MARMOSETS, GOELDI'S MONKEY, TAMARINS AND LION TAMARINS · Callitrichidae

Northern Pygmy Marmoset *Cebuella pygmaea* VU
PT **Mico-leãozinho**, ES **Leoncito**

BR, EC, CO, PE; W Amazonia.
H Lowland humid forest in *terra firme* (but near streams), periodically inundated floodplains, bamboo thickets and secondary growth. Prefers large trees providing gums. N of the rios Solimões/Amazonas and Napo, S of the rios Japurá/Caquetá and Orteguaza.

ID Speckled greenish-brown above, buffy-brownish below; tail upper surface barred black; ears hidden by long backswept hairs forming a mane.

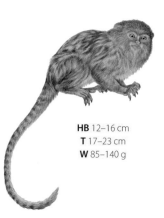

HB 12–16 cm
T 17–23 cm
W 85–140 g

Southern Pygmy Marmoset *Cebuella niveiventris* VU
PT **Mico-leãozinho**, ES **Leoncito**

BO, BR, PE; W Amazonia.
H Lowland humid, *terra firme* forest (near streams), periodically inundated floodplains, bamboo thickets, and secondary growth. Prefers large trees providing gums. S of the rios Solimões/Amazonas and Napo, W of the R. Madeira, N of the R. Madre de Dios.

ID Speckled greenish-brown above; white/off-white on chest and abdomen, variable in extent; tail upper surface barred black; ears hidden by long backswept hairs forming a mane.

HB 12–16 cm
T 17–23 cm
W 85–140 g

Black-crowned Dwarf Marmoset *Callibella humilis* LC
PT **Sauim-anão**

BR **E**; C Amazonia.
H Disturbed lowland primary or secondary *terra firme* forest, white sand (*campinarana*) and seasonally inundated forest (*igapó*). Between the rios Aripuanã and Manicoré, S along the W bank of the R. Roosevelt.

ID Dark olive-brown above; underside orange-yellow to golden greyish-yellow; black crown; white 'eyebrows'; face unpigmented around nostrils; bare ears with tufts in the centre; dark brown tail.

HB 16–17 cm
T 22–25 cm
W 150–185 g

46

NEOTROPICAL PRIMATES

Silvery Marmoset *Mico argentatus* LC
PT **Souim-argênteo**

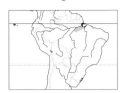

BR **E**; E Amazonia.

H Understorey in secondary (successional) and disturbed primary lowland forest, edge habitat, and forest patches in sandy-soil Amazonian savannah in the alluvial plain S of the R. Amazonas between the lower rios Tocantins and Tapajós.

ID Pale silvery-grey above; creamy-yellow below; face and ears naked and unpigmented pinkish-red; black tail.

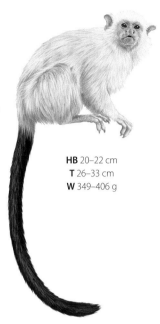

HB 20–22 cm
T 26–33 cm
W 349–406 g

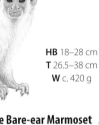

HB 18–28 cm
T 26.5–38 cm
W c. 420 g

Golden-white Bare-ear Marmoset *Mico leucippe* LC
PT **Souim-branco**

BR **E**; E Amazonia.

H Dense vegetation in secondary (successional) and disturbed primary lowland forest, forest patches in savannah enclaves. S of the R. Cuparí, E of the R. Tapajós, S along the rios Jamanxim–Nôvo to the R. São Bento, affluent of the R. Teles Pires.

ID Creamy-white on body and tail; variably pale orange or pale gold on feet, hand, shanks, and tail; face, ears, and genitalia bare, pinkish-scarlet.

HB 19.8–23 cm
T 30.8 cm
W 340–416 g

Snethlage's Marmoset *Mico emiliae* LC
PT **Souim-de-Emília**

BR **E**; E Amazonia.

H Secondary succession and disturbed primary lowland forest, forest patches in savannah enclaves in the S of its range. Between the rios Iriri and Xingu, S to the left bank of the R. Peixoto de Azevedo, a right bank tributary of the R. Teles Pires.

ID Silvery grey, darkening to orange-brown on rump and outer thighs; hands and feet brownish-grey; black patch on crown; tail black.

47

Munduruku Marmoset Mico munduruku VU
PT **Souim-dos-Munduruku, Choim, Suim**

BR **E** ; E Amazonia.

H Lowland primary and secondary (successional) *terra firme* forests. Between the rios Tapajós and Jamanxim–Nôvo, S to the R. Cururú.

ID Mantle, crown, arms, hands, feet, and tail white; lower back and hindlimb beige-yellowish. Face sparsely haired white and skin of face and ears vivid red.

HB 24 cm
T 30 cm
W c. 332 g

No measurements available

Schneider's Marmoset Mico schneideri EN
PT **Souim-de-Schneider**

BR **E** ; E Amazonia.

H Primary and secondary *terra firme* forests and, in the S, forest patches in Cerrado (woodland savannah) forest transition. Between the rios Juruena and Teles Pires, S to their headwaters.

ID Head and upper chest white; crown black; face and ears pink; sparsely haired white; saddle, rump, outer thighs pale lead; underparts cream-silvery, with orange hues; inner forearms, hindlimbs, and feet orangey, tail black except at base.

HB 18–21 cm
T 31–33 cm
W c. 475 g

Black-and-white Tassel-ear Marmoset
Mico humeralifer NT
PT **Souim-de-Santarém**

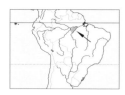

BR **E** ; C Amazonia.

H Disturbed primary forest and secondary *terra firme* forest. Between the rios Maués and Tapajós.

ID Back marbled black with white flecks; underside sparsely furred orange; chest whitish; crown and forehead blackish; long silvery-grey ear tufts; shoulders and arms grey-brown; legs brownish-black; pale hip-patch and thigh-stripe; tail black, ringed.

NEOTROPICAL PRIMATES

Golden-white Tassel-ear Marmoset
Mico chrysoleucos LC
PT **Souim-dourado-e-branco**

BR E ; C Amazonia.

H Disturbed lowland primary and secondary forest. Between the rios Madeira and lower Aripuanã and R. Canumã, S at least to the R. Maracanã.

ID Yellowish-white, with pale golden or orange extremities and underparts; tail faintly banded with darker shade of gold; head always white, with no crown patch; face and ears pink; ear-tufts prominent, white, and fan-like.

HB 20–24 cm
T 32–40 cm
W 280–310 g

HB 20–23 cm
T 34–38 cm
W 315–400 g

Maués Marmoset *Mico mauesi* LC
PT **Souim-de-Maués**

BR E ; C Amazonia.

H Disturbed primary and secondary lowland forest. E of the rios Urariá and Abacaxis to the R. Maués-Açu, S between the rios Tapajós and Sucunduri.

ID Back marbled dark brown and white; neat, trimmed, upstanding ear tufts; face pinkish; light grey cheek patches; crown dark brown to between eyes; pale hip patch; arms and legs dark brown, hands and feet darker; tail black.

HB c. 20 cm
T c. 37 cm
W c. 430 g

Sateré Marmoset *Mico saterei* LC
PT **Souim-de-Sateré**

BR E ; C Amazonia.

H Primary and secondary *terra firme* forest and *igapó* (seasonally inundated forest) between the rios Canumã and Abacaxis, S along the R. Sucunduri.

ID Dark reddish-brown on back, tail, hands, and feet; contrasting, bright whitish-yellow or golden-orange mantle and underparts; nape silvery; grey crown; face and ears bare, pink; genitalia bright orange with unusual lateral pendular skin appendages.

Rio Acarí Marmoset *Mico acariensis* LC
PT **Souim-do-rio-Acari**

BR **E**; C Amazonia.

H Primary and secondary lowland forest between the rios Acarí, Canumã and Sucundurí, E bank affluents of the R. Madeira.

ID Back and thighs dark greyish-brown; belly orange; face bare, mottled orange and brown, framed white; white ruff and upper arms; pale grey cap on crown; lower arms, hands pale orange; lower legs, feet darker orange; pale orange hip-patch and thigh stripe; tail black, dark orange base.

HB c. 22 cm
T c. 38 cm
W c. 350 g

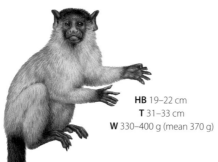

HB 19–22 cm
T 31–33 cm
W 330–400 g (mean 370 g)

Black-headed Marmoset *Mico nigriceps* NT
PT **Souim-de-cabeça-preta**

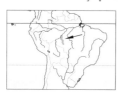

BR **E**; C Amazonia.

H Secondary and disturbed primary lowland forest and forest edge between the R. dos Marmelos, R. Madeira and R. Jiparaná, S and E to the upper R. Roosevelt.

ID Crown, nape, face, ears (bare), hands, feet black; mantle pale greyish-brown; lower back darkens to rump and tail-base; neck, chest pale cream/silvery; arms pale yellow-orange; legs darker orange with pale thigh stripe.

HB c. 22 cm
T c. 36 cm
W 280–310 g

Marca's Marmoset *Mico marcai* NT
PT **Souim-de-Marca**

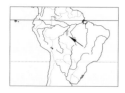

BR **E**; C Amazonia.

H Disturbed primary and secondary lowland forest between the rios Aripuanã and Manicoré, S bank tributaries of the R. Madeira, S at least to the mouth of the R. Roosevelt.

ID Crown dark chestnut; white spot on forehead; pale band around pinkish, bare face; nape and mantle dark grey; lower back pale grey (mottled with dark grey); arms pale and legs darker reddish-brown; tail black.

NEOTROPICAL PRIMATES

Rio Aripuanã Marmoset *Mico intermedius* LC
PT **Souim-do-rio-Aripuanã**

BR **E**; C Amazonia.

H Disturbed primary and secondary (successional) lowland rainforest between the rios Roosevelt and Aripuanã, S probably to their headwaters.

HB 20–24 cm
T 32–40 cm
W 280–300 g

ID Forequarters whitish-silvery; underparts pale orange; rump and tail-base chestnut-brown; legs orangey with pale thigh stripe; tail off-white; naked, pinkish face and ears with sparse white ear-tufts; crown pale grey.

Black-tailed Marmoset *Mico melanurus* NT
PT **Souim-de-rabo-preto**

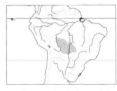

BO, BR, PY; S Amazonia, Chaco, Pantanal margin.

H Lowland secondary and primary forest and dry, vine, and scrub forest, flooded savannah. In CS Brazil, E Bolivia, and sub-humid Chaco in NW Paraguay.

ID Crown, forehead, lower back, and legs dark brown; distinct pale stripe along thigh; mantle drab brown; chest and neck creamy; black tail; face and ears naked, black; face sometimes patchily depigmented.

HB 22–24 cm
T 30–34 cm
W c. 330 g

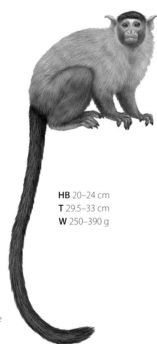

HB 20–24 cm
T 29.5–33 cm
W 250–390 g

Rondon's Marmoset *Mico rondoni* VU
PT **Souim-de-Rondônia**

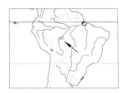

BR **E**; S Amazonia.

H Dense understorey in disturbed primary and secondary forest around tree falls and areas disturbed by human activities between the rios Mamoré, Madeira and Jiparaná, S to the Serra dos Pacaás Novos.

ID Silvery-greyish mantle and chest; lower back darker; muzzle and ears naked, pinkish; white spot on forehead; crown, face, limbs, hands, and feet dark; legs reddish-brown; tail black.

Goeldi's Monkey *Callimico goeldii* VU
PT **Mico-de-Goeldi**, ES **Chichico negro, Supay pichico**

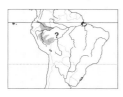

BO, BR, CO, PE; W Amazonia.

H Low in the dense understorey of secondary and lowland *terra firme* bamboo forest, from the R. Caquetá, S Colombia, S through E Peru and W Amazonian Brazil to the Pando of N Bolivia.

ID Long, dishevelled, coal-black fur, broken only by few buff-coloured marks on nape; cape on head and shoulders; head rounded; nose pug-like; face, ears, hands, and feet black.

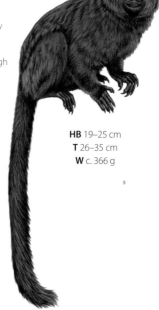

HB 19–25 cm
T 26–35 cm
W c. 366 g

HB 16–21 cm
T 24–31 cm
W c. 322 g

White-tufted-ear Marmoset
Callithrix jacchus LC
PT **Sagüi-de-tufos-brancos, Sagüi-do-nordeste**

BR E; Atlantic Forest, Caatinga.

H Mesophytic and liana forest, *Mauritia* palm woodland, gallery forest, coastal scrub, mangroves, dry deciduous and dry forest (*Caatinga*), *Orbygnia* palm forest on SE border of Amazonia. Introduced in many areas outside its range (e.g., Rio de Janeiro).

ID Back grey-yellowish with transverse stripes; underparts grey; head dark brown to black; white spot on forehead; white fan-shaped ear-tufts; tail grey, ringed.

HB 20–23 cm
T 29–33 cm
W c. 225 g

Black-tufted-ear Marmoset *Callithrix penicillata* LC
PT **Mico-estrela**

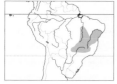

BR E; Cerrado, Caatinga.

H Semi-deciduous and liana forest, *Mauritia* palm woodland (*vereda*), gallery forest in EC Brazil, dry forest in semi-desert scrub and woodland in the NE. Introduced in many areas outside its range.

ID Back, pale brownish-grey to light brown, faint transverse striations; crown, mantle, chest black; pale cheeks; white spot on forehead; long, black, downward-pointing ear tufts; tail dark grey to brownish, ringed.

Wied's Black-tufted-ear Marmoset *Callithrix kuhlii* VU
PT **Sagüi-de-Wied**

HB 20–23 cm
T 29–33 cm
W 350–400 g

BR **E**; Atlantic Forest.

H Coastal evergreen sandy-soil forest (*restinga*) and, inland, humid and mesophytic forest, and wooded cacao plantations between the rios de Contas and Jequitinhonha in S Bahia state.

ID Back black, with tawny hairs on lower back and legs; crown grey; white patch on forehead; whitish cheeks and throat; black tufts in front of ears; hands and feet black or dark brown; tail black, ringed.

△ 0–700 m

HB 18–23 cm
T c. 29 cm
W 230–350 g

White-faced Marmoset *Callithrix geoffroyi* LC
PT **Sagüi-da-cara-branca**

HB 22–25 cm
T 30–35 cm
W c. 400 g

BR **E**; Atlantic Forest.

H Secondary lowland, evergreen, and semi-deciduous forest. In SE Brazil in Espírito Santo and forested E and NE Minas Gerais, N to the rios Jequitinhonha and Araçuaí, S to near the state border of Espírito Santo and Rio de Janeiro.

ID Back brown marbled orange or tawny; white forehead, cheeks, crown, throat, and chest; elongated black ear tufts; hands and feet dark brown; tail black, ringed.

△ 0–500 m

Buffy-headed Marmoset *Callithrix flaviceps* CR
PT **Sagüi-da serra**

BR **E**; Atlantic Forest.

H Disturbed primary and secondary montane and submontane evergreen and semi-deciduous forest in SE Brazil in the Serra da Mantiqueira, S of the R. Doce in Espírito Santo, W into E Minas Gerais.

ID Back yellowish grey-brown, with bases of hairs black with transverse striations; underparts golden-orange; crown, temples, face, and ear tufts ochre; muzzle white; tail buffy greyish, ringed.

△ 270–1800 m

Buffy-tufted-ear Marmoset *Callithrix aurita* EN
PT **Sagüi-da-serra-escuro**

BR **E**; Atlantic Forest.

H Montane evergreen, semi-deciduous, secondary and scrub forest. In SE Brazil in Rio de Janeiro, São Paulo, extending N to the upper R. Doce basin in Minas Gerais.

HB 19–25 cm
T 27–35 cm
W 400–450 g

ID Dark brown/black above with buffy speckling; underside ochre; face, chin, and forehead white to yellowish-buffy; crown whitish to ochre; black ear tufts, with white to dark buffy hairs from inside ear; tail black.

△ 600–1200 m

HB 21–25 cm
T 31–35 cm
W 420–500 g

Spix's Black-mantled Tamarin
Leontocebus nigricollis nigricollis LC

PT **Sauim-de-manto-preto**,
ES **Pichico barba blanca, Bebeleche**

BR, CO, PE; W Amazonia.

H Lowland humid forest, and seasonally flooded and *Mauritia* palm swamp between the R. Içá/Putumayo and the rios Solimões/Amazonas and Napo.

ID Head, neck, mantle, throat, chest, and arms blackish; mantle tapering to faintly striated reddish and black on lower back; rump, legs, tail-base reddish-brown; pale brown around forehead and cheeks; facial skin black; greyish-white hairs on muzzle; tail black.

HB 21–25 cm
T 31–35 cm
W 420–500 g

Graells's Black-mantled Tamarin
Leontocebus nigricollis graellsi NT

ES **Tamarín de dorso negro, Chichico del Napo**

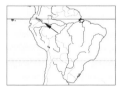

CO, EC, PE; W Amazonia.

H Lowland humid forest, and seasonally flooded and *Mauritia* palm swamp, S of the R. Caquetá, to the N bank of the Napo, E between the ríos Napo and Putumayo to the R. Tamboryacu.

ID Head, neck, mantle, throat, chest, and arms blackish-brown, ticked with buffy hairs; lower back, rump, legs, tail-base dark brown; pale brown forehead and cheeks; facial skin black; grey-white hairs on muzzle; tail black.

Hernández-Camacho's Black-mantled Tamarin
Leontocebus nigricollis hernandezi LC

ES **Bebeleche, Boquiblanco**

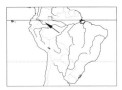

CO **E**; W Amazonia.

H Markedly seasonal lowland and palm forest, shrubby secondary growth, N of the R. Caquetá, W of the R. Yari, in the Orteguaza basin to the R. Caguán.

HB 21–25 cm
T 31–35 cm
W 420–500 g

ID Nape, mantle, throat blackish; mantle tapers down back as stripe to black tail; sides of lower back marbled orange-brown; neck, chest, limbs, rump, sides of body orange-agouti; sides of crown, face pale; whitish hairs on muzzle.

HB 22–32 cm
T 34–35 cm
W 350–400 g

Lesson's Saddle-back Tamarin *Leontocebus fuscus* LC

PT **Sauim-de-costas-malhadas-de-Lesson**, ES **Bebeleche**

BR, CO; W Amazonia.

H Primary and secondary lowland forest between the ríos Japurá/Caquetá and Içá/Putumayo, N of the R. Solimões.

ID Mantle orange-brown, ticked black; saddle, arms, rump, thighs marbled buffy and black; crown, cheeks, facial skin black; grey on muzzle; sides of head reddish-blackish brown; hands and feet blackish mixed with orange; throat, chest, belly rufous; tail black, rufous under base.

HB 22–26 cm
T 32–34 cm
W c. 400 g

Golden-mantled Saddle-back Tamarin *Leontocebus tripartitus* NT

ES **Chichico dorado, Pichico dorado**

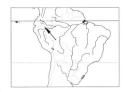

EC, PE; W Amazonia.

H Primary and secondary lowland *terra firme* forest and *Mauritia* palm swamp between the ríos Napo and Curaray, W to the basins of the ríos Yasuní and Nashiño.

ID Mantle golden-orange to creamy; back marbled black, grey, and golden; head, facial skin black; grey-white muzzle; off-white chevron on forehead; arms orange-brown; thighs browner grizzled orange; chest, belly, and inside limbs orange; tail black.

Red-mantled Saddle-back Tamarin
Leontocebus lagonotus LC

ES **Chichico rojo, Pichico común**

HB 22–24 cm
T 30–32 cm
W 350–400 g

EC, PE; W Amazonia.

H Primary and secondary lowland rain forest, S of the ríos Napo and Curaray, S to the ríos Santiago or Morona, Peru.

ID Mantle, rump, outer thighs reddish to dark mahogany; saddle striated black and buffy; crown, forehead, throat, sides of head black; facial skin black; whitish muzzle; pale chevron on forehead; hands and feet black; chest and arms reddish-black or black; tail black, reddish at base.

HB 20–23 cm
T 30–33 cm
W 350–400 g

Andean Saddle-back Tamarin
Leontocebus leucogenys LC

ES **Pichico común, Pichico andino**

HB c. 20 cm
T c. 31 cm
W c. 293 g

PE **E**; W Amazonia.

H Primary and secondary forest, forest edge, riparian forest. E of the Andes, S of the R. Perene to the upper R. Ucayali.

ID Mantle, arms, throat, upper part of chest black or blackish brown; saddle marbled buffy and black; dark patch on orangey-brown thigh; rump orange; hands, feet black; lower chest, belly, and inner thighs reddish, washed with black; tail black, orangey-brown at base.

△ 100–1000 m

Illiger's Saddle-back Tamarin *Leontocebus illigeri* NT

ES **Pichico común**

PE **E**; W Amazonia.

H Primary and secondary lowland and seasonally flooded white-water forest (*várzea*) between the rios Huallaga and Ucayali and lower Tapiche.

ID Mantle (short), outer arms chestnut; forehead and crown black; saddle marbled black and grey or buffy; facial skin black; whitish muzzle; rump and thighs reddish-orange; hands and feet black; chest, belly, and inner limbs reddish; tail black, base reddish.

Geoffroy's Saddle-back Tamarin Leontocebus nigrifrons LC
ES **Pichico común**

PE **E**; W Amazonia.

H Primary and secondary lowland forest between the ríos Amazonas and Yavarí, W to the ríos Ucayali and Tapiche.

HB c. 21 cm
T c. 32 cm
W c. 366 g

ID Mantle greyish-brown; saddle marbled black/greyish; forehead, sides of head black; crown buffy-orange; face skin black; whitish muzzle; upper arms darker; mantle, rump, and thighs reddish-orangey; chest and forearms dark brown to black with orange or reddish hairs; hands, feet, tail black.

HB c. 21 cm
T c. 32 cm
W c. 333 g

Spix's Saddle-back Tamarin
Leontocebus fuscicollis LC

PT **Sauim-de-costas-malhadas-de-Lesson**,
ES **Pichico común**

BR, PE; W Amazonia.

H Primary and secondary *terra firme* forest, S of the R. Solimões, between the ríos Javarí and Juruá, W of the R. Yavarí to the Río Tapiche in Peru.

HB c. 25 cm
T c. 30 cm
W c. 390 g

ID Mantle dark agouti to blackish-brown; back marbled black, with buffy or orange hairs; arms and thighs black ticked with reddish-brown; crown, forehead, temples pale orangey; cheeks, neck, chest dark brown; whitish muzzle; hands and feet black; tail black.

Cruz Lima's Saddle-back Tamarin Leontocebus cruzlimai LC
PT **Sauim-de-Cruz-Lima**

BR **E**; W Amazonia.

H Primary and secondary forest between the rios Teuini and Inauini, left bank affluents of the middle R. Purus.

ID Mantle, crown, underparts, and limbs rusty-orange; back marbled buffy on blackish-brown; rump paler, more yellowish; white chevron on forehead to outer canthus of eye; blackish fur on sides of face and neck; white muzzle; hands and feet blackish; tail black, base reddish.

Ávila-Pires's Saddle-back Tamarin
Leontocebus avilapiresi LC
PT **Sauim-de-costas-malhadas-de-Ávila-Pires**

BR E ; W Amazonia.

H Primary and secondary forest, between the rios Juruá and Purus, possibly S to the R. Tapauá, affluent of the R. Purus.

ID Mantle, rump, forehead, crown, upper arms, thighs blackish-brown finely ticked with orange; neck, chest, belly brown and ticked with orange; lower arms and legs blacker; facial skin black; whitish muzzle; lower legs darker than thighs; hands and feet black; tail black, brown at base.

HB c. 24 cm
T c. 31 cm
W c. 330 g

HB c. 24 cm
T c. 32 cm
W c. 312 g

Grey-fronted Saddle-back Tamarin
Leontocebus mura NT
PT **Sauim-de-costas-malhadas**

BR E ; W Amazonia.

H Primary and secondary forest between the rios Madeira and Purus, probably S to the R. Igapó-açú.

ID Mantle, forehead, crown, and chest dark brown with sparse greyish hairs; saddle strongly marbled ochraceous dark brown to black; greyish around face; whitish muzzle; arms black ticked reddish-brown; thighs, rump, underparts, and base of tail reddish-brown ticked black; hands, feet, and tail black.

HB c. 28 cm
T c. 310
W c. 330 g

Hershkovitz's Saddle-back Tamarin *Leontocebus primitivus* DD
PT **Sauim-de-costas-malhadas-de-Hershkovitz**

BR E ; W Amazonia.

H Primary and secondary forest between the rios Cuniuá and Pauini, affluents of the R. Purus.

ID Mantle, crown, arms, legs, rump, and chest brown, agouti-coloured; distinct pale grey chevron on forehead, separated from crown by blackish line; saddle weakly defined, more reddish; chin grey; throat and neck blackish; upper surface of hands and feet dark agouti; tail blackish, agouti at base.

NEOTROPICAL PRIMATES

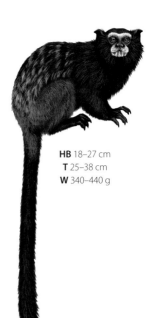

HB 18–27 cm
T 25–38 cm
W 340–440 g

Weddell's Saddle-back Tamarin
Leontocebus weddelli LC

PT **Sauim-de-costas-malhadas-de-Weddell**
ES **Leoncito, Chichilo común**

BO, BR, PE; W Amazonia.

H Primary and secondary forest with dense understories (preference for the latter). Brazil between the rios Purus and Madeira, SE Peru and N Bolivia, E to the R. Beni.

ID Mantle, crown, sides of head, neck, chest, hands, feet, arms black or dark brown; saddle black, marbled buff; white chevron on forehead; facial skin black; muzzle greyish; rump, thighs, underparts reddish-orange; tail black.

HB 18–27 cm
T 25–38 cm
W 340–440 g

White Saddle-back Tamarin
Leontocebus melanoleucus LC

PT **Sauim-branco**, ES **Pichico blanco**

BR, PE; W Amazonia.

H Primary and secondary forest with dense understories between the rios Juruá and Tarauacá, S from the mouth of the R. Eirú just into Peru to the R. Breu.

ID Creamy white, often washed or streaked with yellowish buffy hairs; poorly defined white band on forehead; black ears, facial skin, and external genitalia; muzzle greyish; underparts whitish or yellowish; hands and feet silvery-buff.

HB c. 24 cm
T c. 38 cm
W 360–650 g

Spix's Moustached Tamarin *Tamarinus mystax* LC
PT **Sauim-de-bigode**, ES **Pichico barba blanca**

BR, PE; W Amazonia.

H Primary and secondary lowland forest. S of the R. Solimões, W of the R. Juruá, E to the ríos Tapiche and Ucayali, S to the R. Sheshea.

ID Mantle, shoulders, back, rump, thighs upper arms blackish-brown with orangey ticking; pink muzzle; white moustache; hairs on chin black; crown, arms, legs, hands, feet, tail black; throat, chest, underparts blackish-brown; external genitalia pink with white hairs.

Red-capped Moustached Tamarin
Tamarinus pileatus pileatus LC
PT **Sauim-de-bigode**

HB c. 24 cm
T c. 38 cm
W 360–650 g

BR E; W Amazonia.

H Primary and secondary lowland rainforest. W of the R. Purus, S to the rios Pauiní or Mamoria.

ID Mantle blackish-brown with buffy or orange ticking; crown and forehead to thin line between eyes rusty-red; face black; pink muzzle; white moustache; back and rump mixed blackish-brown and buffy hairs; arms and thighs brown; hands and feet blacker; tail black; external genitalia pink with white hairs.

White-rumped Moustached Tamarin
Tamarinus pileatus pluto LC
PT **Sauim-de-bigode**

HB c. 24 cm
T c. 38 cm
W 360–650 g

BR E; W Amazonia.

H Primary and secondary lowland forest. W of the R. Purus, N of the R. Tapauá, W to the R. Coarí.

ID Like *T. mystax* except pink, white-haired anogenital-inguinal region and ventral tail-base. Mantle black, ticked buff; crown, head, arms, hands, feet, tail black; back and thighs like mantle but buffier; muzzle pink; white moustache; black patch on chin; throat, chest, underparts blackish-brown.

Kulina's Moustached Tamarin *Tamarinus kulina* NE
PT **Sauim-de-bigode**

BR E; W Amazonia.

H Primary and secondary lowland rainforest, S of the R. Amazonas, E of the lower R. Juruá, west to the R. Tefé.

ID Mantle and forelimbs blackish-brown, hairs with subterminal yellow band, terminal band black; crown, saddle, rump, hindlimbs dark blackish-brown; white moustache; black tail.

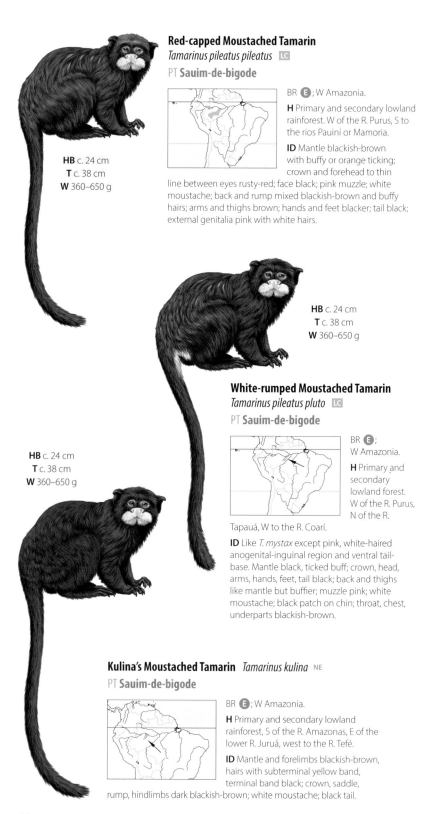

Geoffroy's Red-bellied Tamarin
Tamarinus labiatus labiatus LC

PT **Sauim-de-bigode**,
ES **Pichico de barriga anaranjada**

HB 21–28 cm
T 30–38 cm
W 400–510 g

BO, BR, PE; W Amazonia.

H Primary and secondary evergreen and semi-deciduous forest between the rios Purus and Madeira–Abunã, S of the R. Ipixuna, S to the R. Tahuamanú in N Bolivia, just entering SE Peru.

ID Nape and mantle blackish-brown; back black, marbled white; reddish-orangey underparts; throat black; white crown; facial skin black; white moustache outlining mouth; tail black, red or orangey under base.

Gray's Red-bellied Tamarin
Tamarinus labiatus rufiventer LC

PT **Sauim-de-bigode**

HB 21–28 cm
T 30–38 cm
W 400–510 g

BR **E**; W Amazonia.

H Primary and secondary forest between the rios Madeira and Purus, S to the R. Ipixuna.

ID Like *S. l. labiatus*, but with red, Y-shaped mark on front of crown and slight silvery patch behind. Nape and mantle blackish-brown; back black, marbled white; reddish-orangey underparts; throat and facial skin black; white moustache around mouth; tail black, reddish-orangey under base.

Thomas's Red-bellied Tamarin *Tamarinus thomasi* LC
PT **Sauim-de-bigode**

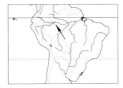

BR **E**; W Amazonia.

H Primary and secondary forest between the rios Japurá and Solimões, from the Auati-Paraná, W to the R. Tonantins.

ID Nape and mantle blackish; back black, marbled white; orange (not reddish) underparts; throat, upper chest black; crown black, with pale silvery mid-line; white moustache outlining mouth; tail black, red or orangey under base.

Black-chinned Emperor Tamarin   *Tamarinus imperator*   LC
PT **Bigodeiro**, ES **Pichico emperador**

BR, PE; W Amazonia.

H Primary and secondary lowland rainforest. E of the upper R. Purus, between the rios Purus and Acre just into Peru.

ID Body greyish-agouti; rusty-red or orange on underside; inner arms silvery orange; crown silvery-brown; facial skin black; elongated white moustaches, hanging down to forearms; short white chin hairs; black chin patch; hands and feet black; tail orangey-red, with dark tip.

HB 23–26 cm
T 35–42 cm
W 400–550 g

Bearded Emperor Tamarin   *Tamarinus subgrisescens*   LC
PT **Bigodeiro**, ES **Pichico emperador**, **Mono bigotudo**

BO, BR, PE; W Amazonia.

H Primary and secondary forest. E of upper R. Juruá to the R. Tarauacá, W to the R. Urubamba, S to the R. Muyumanu basin in Peru and Bolivia.

ID Body brown-agouti; underside and inner arms grizzled-brown; inner sides of arms silvery brown; facial skin black, elongated white moustaches; chin whiskers form small beard covering black chin patch; hands and feet black; tail orangey-red with dark tip.

HB 23–26 cm
T 35–42 cm
W 400–550 g

HB 21–26 cm
T 33–41 cm
W c. 430 g

Mottled-face Tamarin   *Tamarinus inustus*   LC
PT **Sauim-de-cara-manchada**, ES **Tití diablito**

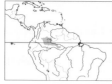

BR, CO; W Amazonia.

H Lowland primary and secondary *terra firme* forest, seasonally flooded forest between the upper R. Negro and R. Japurá/Caquetá, W to the lower R. Yari in Colombia and N through the Apaporis and Vaupés basins.

ID Body and tail black; chocolate-brown on back; cinnamon on flanks; face pink, mottled white; short white hairs on upper lip and nostrils; ears black and bare; white hairs on genitals.

Golden-handed Tamarin *Saguinus midas* LC
PT **Sauim-de-mão-dourada**, FR **Tamarin à mains dorées**, SR **Kusi**

BR, GF, GY, SR; E Amazonia.

H Primary, secondary forest, savannah forest patches, montane savannah, liana, marsh, and swamp forest. N of the R. Amazonas, E of the R. Negro in Brazil, E of the Essequibo R.

HB 24–25 cm
T 38–39 cm
W 380–500 g

ID Back, rump, and thighs black, spotted, ticked with buff; face, head, shoulders, upper arms, tail, underparts black; hands and feet orange or yellowish-orange; ears notched, bare and black.

Western Black-handed Tamarin *Saguinus niger* VU
PT **Sauim-de-mão-preta-do-oeste**

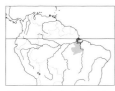

BR **E**; E Amazonia.

H Primary lowland, submontane and secondary forest, swamp, forest edge. S of the R. Amazonas, E of the rios Xingu–Fresco to the R. Tocantins, S to the R. Gradaús. Also, SW part of Marajó I.

ID Similar to *S. midas* but smaller and with larger ears. Fur entirely black, except lower back spotted, ticked with reddish-brown; facial skin, hands, feet black; ears notched, bare, and black.

HB 21–26 cm
T 32–40 cm
W c. 431 g

HB 21–26 cm
T 32–40 cm
W c. 431 g

Eastern Black-handed Tamarin *Saguinus ursula* VU
PT **Sauim-de-mão-preta-do-leste**

BR **E**; E Amazonia.

H Primary lowland, submontane and secondary forest, swamps, forest edge. S of the R. Amazonas, E of the R. Tocantins to forest patches at the transition to Cerrado and Caatinga in Maranhão.

ID Mantle black with long, dishevelled hairs; lower back, rump, thighs boldly striated, marbled golden buffy; facial skin, hands, feet black; ears notched, bare, and black.

Pied Tamarin *Saguinus bicolor* CR
PT **Sauim-de-coleira**

BR E ; C Amazonia.

H Lowland primary and secondary forest (*capoeira*), white sand (*campinarana*), and seasonally flooded forest, E of the R. Negro, N to the R. Cuieiras, E to the R. Urubu.

ID Mantle, arms, feet, neck, chest, crown whitish; abruptly demarcated from pale reddish-brown back, rump, and legs; underside orange or reddish-gold; face and notched ears black; tail dark brown above, orange below.

HB 23–33 cm
T 34–42 cm
W 480–600 g

HB 21–28 cm
T 34–42 cm
W 400–600 g

Martins's Bare-faced Tamarin
Saguinus martinsi martinsi NT
PT **Sauim**

BR E ; C Amazonia.

H Primary and secondary lowland rainforest. N of the R. Amazonas, between the rios Nhamundá and Trombetas, N to the Cachoeira Porteira (Trombetas).

ID Body brownish from crown to tail base; paler on flanks and shoulders; orange underparts; limbs buffy, orange below; face and front of crown almost naked; white hands and feet; blue spots on ears; head puce; no white on ruff or nape.

HB 21–28 cm
T 34–42 cm
W 400–600 g

Ochraceous Bare-faced Tamarin *Saguinus martinsi ochraceus* NT
PT **Sauim**

BR E ; C Amazonia.

H Primary and secondary lowland rainforest between the rios Uatumã and Nhamundá, N of the R. Amazonas.

ID Body paler than *S. m. martinsi*, generally yellowish-brown; golden-orange underparts; silvery to buffy ruff and nape; neck, base of mantle greyish-ochre or yellowish-grey; face, forehead, crown blackish; arms, hands, and feet ochraceous.

NEOTROPICAL PRIMATES

White-footed Tamarin *Oedipomidas leucopus* VU
ES **Tití gris**

CO **E**; N Colombia.

H Primary, secondary lowland, and pre-montane rainforest between the lower R. Cauca and the middle R. Magdalena, S to Mariquita. On Mompos I. in the R. Magdalena.

HB 22–29 cm
T 35–42 cm
W c. 462 g

ID Back brownish-grey with long whitish hair tips; underparts rusty-orange; white brow; long dark brown hairs on nape forming a ruff; facial skin black; silvery hairs on cheeks and forearms; hands, feet whitish; tail brown, tipped white.

HB 21–26 cm
T 33–40 cm
W c. 417 g

Cotton-top Tamarin *Oedipomidas oedipus* CR
ES **Tití cabeza de algodón**

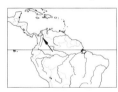

CO **E**; N Colombia.

H Primary and secondary lowland evergreen and dry deciduous forest, scrubland, forest edge between the R. Atrato and lower ríos Cauca and Magdalena, and NE Chocó.

HB 20–29 cm
T 32–42 cm
W c. 486 g

ID Back grey-brown; underparts, forearms, lower legs, hands, feet, and exuberant crown white; variable red patches on thighs, rump, and shoulders; face and bare ears black; short white hairs frame face; tail half red, half dark brown.

△ 0–400 m

Geoffroy's Tamarin *Oedipomidas geoffroyi* NT
ES **Mono tití panameño, Tití**

N CO, PA.

H Primary, secondary, and dry deciduous forest along the Pacific coast, Colombia, S to the R. San Juan, and Panama from Darién to just W of the Canal Zone. Introduced population E of the Azuero Peninsula.

ID Lower back, upper arms, legs dappled black and yellow; underside pale yellow; nape dark reddish; forearms white; wedge-shaped, white crest; face black with sparse white hairs framing cheeks; tail reddish then black.

△ 0–350 m

HB 26–33 cm
T 32–40 cm
W c. 710 g (M),
c. 795 g (F)

Golden Lion Tamarin *Leontopithecus rosalia* EN
PT **Mico-leão-dourado**

BR **E**; Atlantic Forest.

H Primary and secondary coastal lowland forest; sometimes cultivated and secondary regrowth forest. Largely restricted to two municipalities, Silva Jardim and Cabo Frio, in the state of Rio de Janeiro.

ID Body entirely golden-orange; a mane; face bare, pale purplish; orange, brown or black patches on tail and hands.

△ Below 300 m

Golden-headed Lion Tamarin
Leontopithecus chrysomelas EN
PT **Mico-leão-de-cara-dourada,
Mico-leão-baiano**

BR **E**; Atlantic Forest.

H Coastal lowland forest and coastal white-sand forest (*restinga*), abundant in piaçava palms (*Attalea funifera*), secondary forest and old cacao plantations (*cabruca*), between the rios Jequitinhonha and de Contas, in Bahia.

ID Body largely black; thick, long, golden-red and reddish-orange hairs on front of mane, forearms, hands, feet, thighs, and upper side of tail-base.

△ Up to 700 m

HB 22–26 cm
T 33–39 cm
W 540–700 g

NEOTROPICAL PRIMATES

HB 25–30 cm
T 36–41 cm
W 540–690 g,
(male slightly larger than female)

Black Lion Tamarin *Leontopithecus chrysopygus* EN
PT **Mico-leão-preto**

BR E; Atlantic Forest.

H Patches of inland, lowland, mesophytic semi-deciduous forest, and swampy forest, along the N bank of the R. Paranapanema, W to the R. Paraná, and between the upper Paranapanema and R. Tietê, São Paulo.

ID Body mostly black; golden-reddish patches on forehead, rump, base of tail, thighs, and ankles; long mane with some red-gold hairs.

△ 300–790 m

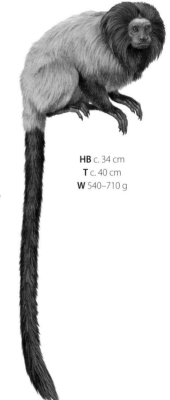

Black-faced Lion Tamarin *Leontopithecus caissara* EN
PT **Mico-leão-de-cara-preta, Mico-leão-caiçara**

BR E; Atlantic Forest.

H Primary and secondary lowland coastal forest on the São Paulo state mainland, sandy-soil forest (*restinga*), and swampy and inundated forest on Superagüi I. in the state of Paraná. The most southerly of the callitrichids.

ID Body orange-gold; mane, face, chest, feet, forearms, and tail black; transitional region behind shoulders with long golden hairs, dark brown at base.

△ 0–40 m

HB c. 34 cm
T c. 40 cm
W 540–710 g

67

SQUIRREL MONKEYS, GRACILE CAPUCHINS AND ROBUST CAPUCHINS · Cebidae

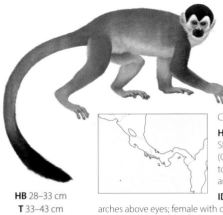

Black-crowned Central American Squirrel Monkey
Saimiri oerstedii oerstedii EN
ES **Mono tití, Mono tití chiricano**

CR, PA; C America.

H Lowland evergreen forest, on Pacific coast SE Costa Rica (Puntarenas) and SW Panama (Chiriquí), from S bank of R. Grande de Térraba to the mouth of the R. Chiriquí Nuevo and the archipelago in the Golfo de Chiriquí.

HB 28–33 cm
T 33–43 cm
W 750–950 g (M),
600–800 g (F)

ID Black crown; tufted ears; white 'Gothic' type arches above eyes; female with dark sideburns; legs and arms orange; back and sides bright or reddish-orange.
△ Up to 1450 m

Grey-crowned Central American Squirrel Monkey
Saimiri oerstedii citrinellus EN
ES **Mono tití**

CR **E**; C America.
H Lowland evergreen forest, Pacific coast W Costa Rica in Puntarenas Province, NE limit R. Tulín, S limit R. Grande de Térraba.

ID Crown agouti in male, black in female; tufted ears; white 'Gothic' type arches above eyes; female with black sideburns; back and flanks largely bright orange or reddish-orange; outer sides of legs buffy or greyish-agouti.
△ Up to 500 m

HB 28–33 cm
T 33–43 cm
W 750–950 g (M),
600–800 g (F)

Humboldt's Squirrel Monkey
Saimiri cassiquiarensis LC
PT **Macaco-de-cheiro**, ES **Mono ardilla**

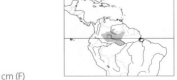

BR, CO, VE; N Amazonia.

H Humid forests, W of the R. Negro, S to the *várzea* of the lower Japurá–Solimões interfluvium, then W, N of the Japurá, N into Venezuela and W into Colombia, N of the R. Apaporis to the Chiribiquete Plateau.

HB 25–37 cm (M), 28–34 cm (F)
T 38–45 cm (M), 36–43 cm (F)
W 650–1125 g (M), 550–1200 g (F)

ID Yellow tone to grey crown; bright reddish-tan on back; weakly contrasting, light nuchal collar; forearms orange.

NEOTROPICAL PRIMATES

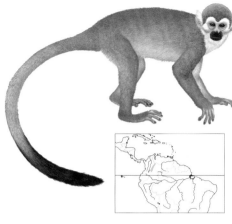

HB 25–37 cm (M), 28–34 cm (F)
T 38–45 cm (M), 36–43 cm (F)
W 650–1125 g (M), 550–1200 g (F)

Colombian Squirrel Monkey
Saimiri albigena VU
ES **Mono ardilla**

CO **E** ; E Llanos.

H Eastern Andes of Colombia in low-canopy gallery forests of the Eastern Llanos, foothills of the Cordillera Oriental, and palm swamp forest.

ID Similar to *S. cassiquiarensis*, but with mainly greyish-agouti forearms and hands, and orange back; crown and nape greyish or buffy-agouti; tufted ears.

Ecuadorian Squirrel Monkey *Saimiri macrodon* LC
PT **Macaco-de-cheiro**, ES **Mono ardilla, Barizo, Fraile**

BR, CO, EC, PE; W Amazonia.

H Seasonally inundated forest, extending W from the rios Juruá and Japurá into S Colombia to the R. Apaporis, to E Ecuador, and N Peru.

ID Crown greyish olivaceous, washed with orange; white 'Gothic' type arches above eyes; muzzle black or grey; ears white, slightly furred and pointed; inside of legs and arms pale; feet and hands deep yellowy orange; tail pencil black.
△ 200–1200 m

HB 25–32 cm
T 34–44 cm
W 835–1380 g (M) 590–1150 g (F)

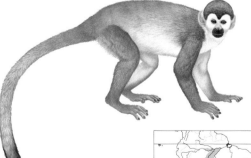

Golden-backed Squirrel Monkey
Saimiri ustus NT
PT **Macaco-de-cheiro**

BR **E** ; C Amazonia.

H *Terra firme* forest, flooded forest. S of the R. Solimões/Amazonas, from R. Tefé E to R. Xingu–Iriri, and S to upper R. Guaporé and R. Juruena. Range limits poorly known.

HB 25–35 cm (M), 23–42 cm (F)
T 40–45 cm (M), 31–42 cm (F)
W 710–1200 g (M), 620–880 g (F)

ID Naked face; bare ears; 'Gothic' type arches; crown agouti (or black in female); golden back; thighs greyish-agouti; hands and feet orange, sometimes to forearms.

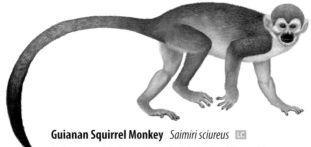

HB 25–37 cm (M),
25–34 cm (F)
T 36–40 cm (M),
36–47 cm (F)
W 550–1400 g (M),
550–1200 g (F)

Guianan Squirrel Monkey *Saimiri sciureus* LC
PT **Macaco-de-cheiro**, FR **Saimiri**, SR **Monki-monki**

BR, GF, GY, SR; N & E Amazonia.

H Lowland forest, seasonally inundated forest, river edge, mangrove forest. N of the R. Amazonas, W to the rios Negro and Branco.

ID Agouti crown (mixed black in female); 'Gothic' type arches; white ear tufts; pink face; black muzzle; yellowish-orange forearms, hands, and feet; underside white.

Collins's Squirrel Monkey *Saimiri collinsi* LC
PT **Macaco-de-cheiro**

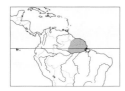

HB 25–37 cm (M), 25–34 cm (F)
T 36–40 cm (M), 36–47 cm (F)
W 550–1400 g (M), 550–1200 g (F)

BR **E**; E Amazonia.

H Lowland flooded forests, river edge, mangroves, Amazon–Cerrado transition. N Brazil, Marajó I. (R. Amazonas estuary), S Pará state, E of R. Tapajos, N Tocantins to Maranhão states.

ID Grey and yellow crown; 'Gothic' type arches; white ear tufts; pink face; black muzzle; head, shoulders, and back greyish; hands and feet dark rich tawny.

HB 28–31.5 cm (M), 26.5–28.5 cm (F)
T 38–43 cm (M), 39.5–41 cm (F)
W c. 1 kg (M), 700–900 g (F)

Bolivian Squirrel Monkey
Saimiri boliviensis boliviensis LC

PT **Macaco-de-cheiro**, ES **Chichilo, Fraile, Mono ardilla, Huasa**

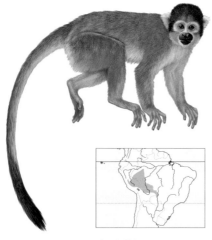

BO, BR, PE; S & W Amazonia.

H Humid forest, river edge, flooded forests. S of the R. Solimões, from the rios Juruá and Tefé, S into SE Peru and N Bolivia W of the R. Guaporé.

ID Blackish crown; 'Roman' arch above eyes; male body grey and female blackish; golden-yellow forearms, hands, and feet; thin black tail pencil.

△ 50–800 m

NEOTROPICAL PRIMATES

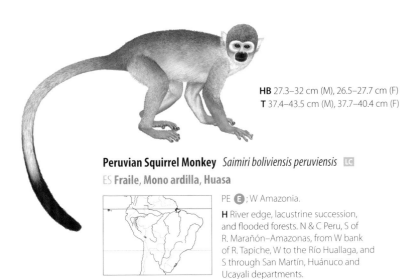

HB 27.3–32 cm (M), 26.5–27.7 cm (F)
T 37.4–43.5 cm (M), 37.7–40.4 cm (F)

Peruvian Squirrel Monkey *Saimiri boliviensis peruviensis* LC
ES **Fraile, Mono ardilla, Huasa**

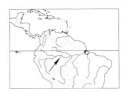

PE **E** ; W Amazonia.

H River edge, lacustrine succession, and flooded forests. N & C Peru, S of R. Marañón–Amazonas, from W bank of R. Tapiche, W to the Río Huallaga, and S through San Martín, Huánuco and Ucayali departments.

ID Crown agouti in male, black or blackish-agouti in female; bright yellow forearms; tail greyish to blackish-agouti above, with black tip.

Black-headed Squirrel Monkey *Saimiri vanzolinii* EN
PT **Macaco-de-cheiro-de-cabeça-preta**

BR **E** ; C Amazonia.

H White-water flooded forest (*várzea*). N of the R. Solimões, in Mamirauá SD Reserve between the Paraná do Jarauá, Paraná do Aiucá in the W, and R. Japurá in the E. Sympatric with *S. cassiquiariensis* or *S. macrodon* in some areas.

ID Black crown; short dense pelage; broad continuous black band from crown to tail tip; shoulders greyish; hands, forearms, feet light burnt yellow.

HB 26–30 cm (M), 22–26 cm (F)
T 35–40 cm (M), 37–41 cm (F)
W c. 950 g (M), c. 650 g (F)

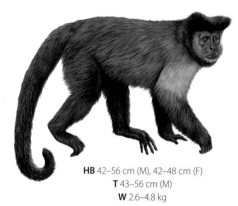

HB 42–56 cm (M), 42–48 cm (F)
T 43–56 cm (M)
W 2.6–4.8 kg

Northern Black-horned Capuchin
Sapajus nigritus NT
PT **Macaco-prego**

BR **E**; Atlantic Forest.

H Gallery and secondary forest. SE Brazil, S of the R. Doce and R. Grande in Minas Gerais and Espírito Santo, E of the Rio Paraná, S to R. Tietê, towards Curitiba. Possibly limited to coastal Rio de Janeiro, inland to the R. Paraíba do Sul with *S. cucullatus* extending N to Minas Gerais and Espírito Santo.

ID Dark crown; horn-like tufts, sometimes contrasting light face; dark brown coat, often reddish or yellow-fawn underparts; tail black.

Southern Black-horned Capuchin
Sapajus cucullatus NT
PT **Macaco-prego**, ES **Caí**

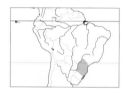

AR, BR; Atlantic Forest.

H Subtropical forest, gallery, and secondary forest. NE Argentina (Misiones Province) and SE Brazil, S of R. Tietê extending S through Atlantic Forest, E of the R. Paraná into N Rio Grande do Sul State, to c. 29° 50′ S. The most S occurring capuchin.

ID Dark crown and horn-like tufts; bright contrasting white border around face; black coat; tail black.

HB 42–56 cm (M), 42–48 cm (F)
T 43–56 cm (M)
W 2.6–4.8 kg

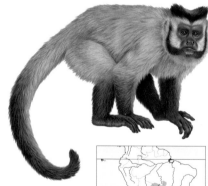

HB 40–45 cm
T 41–47 cm
W 3–3.5 kg

Hooded Capuchin *Sapajus cay* VU
PT **Macaco-prego-do-papo-amarelo**, ES **Caí, Silbador, Mono capuchino de Azara**

AR, BO, BR, PY; Southern Cone.

H Subtropical humid and semi-deciduous forest, gallery and montane forest and Pantanal wetlands. S Bolivia, NW Argentina, E Paraguay, and Mato Grosso, S Goiás, and Mato Grosso do Sul.

ID Crown pale to blackish-brown; small tufts; small, short-limbed; prominent dark dorsal stripe; mostly pale; neck and tail slightly burnt brown; dorsal parts of body greyish-brown; extremities blackish.
△ Up to 1500 m

Crested Capuchin *Sapajus robustus* EN
PT **Macaco-prego-de-crista**

HB 42–56 cm (M), 33–44 cm (F)
T 43–56 cm
W 1.3–4.8 kg

BR **E** ; Atlantic Forest.

H Humid to dry semi-deciduous forest.

S from the R. Jequitinhonha in S Bahia to R. Doce and R. Suaçuí Grande in Espírito Santo, and Minas Gerais. E of Serra do Espinhaço. Hybrids with *S. nigritus* between R. Suaçuí Grande and R. Santo Antônio.

ID Black crown tufts join together at midline; face dark greyish; very dark limbs, extremities; underparts red or yellow.

Bearded Capuchin *Sapajus libidinosus*
PT **Macaco-prego**

BR **E** ; Atlantic Forest, Caatinga, Cerrado.

H C and NE Brazil, W and N of R. São Francisco to Maranhão state, W of Piauí, E to C Rio Grande do Norte, NW Paraíba, W Pernambuco, and W Alagoas, S to Minas Gerais.

ID Black crown; rounded black tufts; dark brown sideburns; orangey-yellow throat and dorsum, flanks, outer parts of arms, and proximal tail; forearms dark; lower back and thighs greyish-brown.

△ 300–600 m

HB 34–44 cm
T 38–49 cm
W 1.3–4.8 kg

Yellow-breasted Capuchin
Sapajus xanthosternos CR
PT **Macaco-prego-de-peito-amarelo, Pichicau**

HB 39–42 cm (M), 36–39 cm (F)
T 38–45 cm
W 2–4.8 kg (M), 1.3–3.4 kg (F)

BR **E** ; Atlantic Forest, Caatinga.

H Humid tropical forest. In W, dry forest. S and E of the R. São Francisco, S to R. Jequitinhonha in the S of Bahia, SW to R. das Velhas, R. Pardo Grande and the Serra do Espinhaço.

ID Black cap; face and temples fawn; usually brindled reddish above; sharply marked, golden-red underside; black tail and limbs. SW populations considerably darker in overall colour.

Blond Capuchin *Sapajus flavius* **EN**
PT **Macaco-prego-galego**

BR **E**; Atlantic Forest, Caatinga.

H Humid coastal forest and dry forest in NE Brazil from Rio Grande do Norte, S through Paraíba, NE Pernambuco to R. São Francisco in Alagoas.

ID White to yellow cap; throat flap in older males; face pinkish; golden-yellow body and limbs; hands and feet black; tail blond.

HB 36–46 cm (M), 32–42 cm (F)
T 37–43.5 cm (M), 35–38 cm (F)
W 1.5–4.2 kg (M), 1.5–2 kg (F)

Guianan Brown Capuchin *Sapajus apella apella* **LC**
PT **Macaco-prego**, ES **Mono capuchino pardo**, FR **Capucin Brun**, SR **Keskesi**

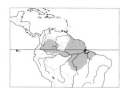

BR, CO, GF, GY, SR, VE; E & N Amazonia.

H Humid forest. E of the rios Madeira and Aripuanã, N of the rios Amazonas/Solimões and Japurá, E to Guianas and S Orinoco Delta. S limits defined by the forest-Cerrado ecotone, extends E to Maranhão.

ID Broad, square head; face flat, light grey-brown; short tufts; short limbs; extremities and tail dark; grey to brown above; yellowish or red underside.

△ Up to 1000 m

HB 38–46 cm
T 38–49 cm
W 2.3–4.8 kg (M), 1.3–3.4 kg (F)

Margarita Island Capuchin
Sapajus apella margaritae **CR**
ES **Mono capuchino pardo**

VE **E**; Caribbean island.

H Pre-montane and montane forest, disturbed palm forest. Isolated forest patches in E of Margarita Island, Venezuela.

ID Darker and smaller than *S. a. apella* forms from upper Orinoco, Venezuela; short crown tufts; greyish face; pink cheeks and chin; upper arms and shoulders pale yellow; extremities black; body mostly pale brown; dorsal stripe.

HB 38–46 cm
T 38–49 cm
W 2.3–4.8 kg (M), 1.3–3.4 kg (F)

Large-headed Capuchin *Sapajus macrocephalus* LC
PT **Macaco-prego**, ES **Maicero**, **Barizo**, **Machin café**, **Mono negro**

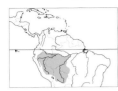

BO, BR, CO, EC, PE; W Amazonia.

H Humid lowland forest. S of the ríos Amazonas/Solimões and Japurá/Caquetá, W of the ríos Madeira and Aripuanã.

ID Crown cap variable in shape; often tall rounded horns; overall grey-brown to dark brown above; dark dorsal stripe; yellow or red-gold below; limbs and tail dark.

△ Up to 1800 m, rarely 2700 m

HB 37.5–45.5 cm (M), 39.5–41 cm (F)
T 42.5–49 cm (M), 41.6–42 cm (F)
W 2.9–4.6 kg (M), 1.3–3.4 kg (F)

Peruvian White-fronted Capuchin *Cebus yuracus* NT
PT **Caiarara**, ES **Machín blanco**, **Capuchino de frente blanca**

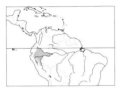

BR, CO, EC, PE; W Amazonia.

H Wet lowland *terra firme*, seasonally inundated forest in upper Amazon Basin, to montane forests on W Andes in S Colombia, E Ecuador, NE Peru, and possibly W Brazil.

ID Cap dark brown; grey-fronted forehead, face, chest, arms; body ochraceous brown; greyish or buffy on outer forelimbs; flanks paler; underparts pale silvery to ochraceous orange.

△ Up to 2000 m

HB c. 43 cm (M), c. 37 cm (F)
T c. 47 cm (M), c. 45 cm (F)
W 2–4.7 kg

HB c. 40 cm (M), 39–46 cm (F)
T c. 44 cm (M), 39–47.5 cm (F)
W 2.8–3 kg

Shock-headed Capuchin
Cebus cuscinus NT
ES **Machín blanco**, **Mono blanco**

BO, PE; SW Amazonia.

H Lowland *terra firme*, seasonally inundated forest. S bank upper R. Purus in SE Peru, W into Cuzco, upper R. Madre de Dios, S and E to Tambopata Basin into NW Bolivia.

ID Cap large, dark brown; body tawny to ochraceous; limbs brown; forearms orange-red; wrists and hands darker; feet brown to auburn; underparts orange, silvery or buff; tail cinnamon brown.

△ Up to 1800 m

Spix's White-fronted Capuchin *Cebus unicolor* VU
PT **Caiarara**, ES **Machín blanco, Mono blanco**

BO, BR, PE; W & C Amazonia.

H Lowland *terra firme*, seasonally inundated forest and Amazonian savannahs. Widespread in upper Amazon Basin, S of R. Negro, W from R. Tapajós, through N of Mato Grosso and Rondônia to R. Ucayali in E Peru. N Bolivia, S to Pando, Beni, and La Paz departments.

HB 36.5–37.5 cm
T 42–46 cm

ID Crown nearly black; no white on shoulders; body bright ochre or greyish-brown, with yellowish or creamy-fawn front; reddish-yellow limbs and tail.

Humboldt's White-fronted Capuchin *Cebus albifrons* LC
PT **Caiarara**, ES **Mono cariblanco, Maicero cariblanco**

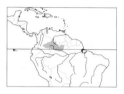

BR, CO, VE; W & C Amazonia.

H Deciduous, mangrove, flooded forest, gallery forest in E Llanos of Colombia. Upper Amazon, S Venezuela, S & E Colombia, and NW Brazil, N of rios Negro and Branco.

HB c. 37.5 cm (M), 36.5–37.5 cm (F)
T c. 42.5 cm (M), 41–46 cm (F)

ID Short, dark, rounded cap; light forehead; face naked, pink; front creamy; body pale greyish-brown; darker orangish on limbs; hands and feet yellowish-brown.

△ Up to 2000 m

HB 37–46 cm
T 45–55 cm
W 3–4.2 kg (M), 2.3–3 kg (F)

Guianan Weeper Capuchin
Cebus olivaceus LC

PT **Caiarara**, ES **Mono capuchino común**

BR, GY, VE; N Amazonia, Llanos, Caribbean coast.

H Primary forest, gallery forest, shrub woodland. Upper R. Orinoco, and savannah, W to Falcon region and the Cordillera de Merida, E to R. Essequibo in W Guyana, S to rios Negro and Branco.

ID Face and forehead light grey-brown; dark V-shaped crown cap; face pink; shaggy fur; pelage dark brown or reddish with black-agouti banding; underside dark; black extremities.

△ Up to 2000 m

Chestnut Weeper Capuchin *Cebus castaneus* LC
PT **Caiarara**, FR **Capucin à tête blanche**, SR **Bergi keskesi**

BR, GF, GY, SR; Guiana Shield.
H Primary rainforest, gallery forest. In Guianas from R. Essequibo, E through interior Suriname and French Guiana, and N Brazil, from E of the R. Branco, S to R. Amazonas, E to the Atlantic coast in Amapá.

HB 37–46 cm
T 45–55 cm
W 3–4.2 kg (M), 2.3–3 kg (F)

ID Narrow black triangle on crown; head yellowish-white to reddish-chestnut; upperparts reddish-chestnut; shoulders and arms pale yellow, hands and feet blackish.

△ Up to 2000 m

Ka'apor Capuchin *Cebus kaapori* CR
PT **Caiarara**

BR 🅔; E Amazonia.
H Tall lowland *terra firme* forest, babassu palm forest in NE Pará and NW Maranhão, R. Tocantins in Pará to right bank R. Grajaú, Maranhão. Possibly Marajó Is.

HB 37–46 cm
T 40–55 cm
W c. 3 kg (M), c. 2.4 kg (F)

ID Crown with triangular black cap; dark stripe down nose; long body, greyish agouti brown; lighter on flanks; face, shoulders, mantle, and tail tip silvery-grey; limbs agouti; hands and feet dark brown or black.

Sierra de Perijá White-fronted Capuchin
Cebus leucocephalus VU
ES **Mono cariblanco**

CO, VE; N Andes.
H Lowland moist forest, semi-deciduous dry forest, mangroves. N Colombia from W slope of Cordillera Oriental in Santander Department, E to Norte de Santander and NW Venezuela.

HB 37–40.7 cm (M)
T 39.2– 49.9 cm (M)

ID Reddest of gracile capuchins; cap cinnamon brown to bistre; back, shoulders, upper arms, and thighs cinnamon-brown; forelegs and front of thighs burnt sienna; extremities auburn; tail cinnamon brown above, buffy underneath.

HB 45–50.5 cm
T 42–45.5 cm

Varied White-fronted Capuchin *Cebus versicolor* EN
ES **Mono cariblanco**

CO **E** ; N Andes.

H Lowland moist forest and palm swamps in middle R. Magdalena Valley. N Colombia in the middle R. Magdalena except for the W slope of the Cordillera Oriental, from the S portion of the Magdalena Department S to the departments of Cundinamarca and Tolima.

ID Dark maroon-brown overall; red tones on mid-dorsal region and foreparts of limbs; little to no white on shoulders and chest.

HB 34.8–40.7 cm (M),
35.3–38.5 cm (F),
T 41.9–49.5 cm (M),
46.1–50 cm (F)

Río Cesar White-fronted Capuchin *Cebus cesarae* EN
ES **Mono cariblanco**

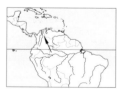

CO **E** ; N Andes.

H Dry semi-deciduous forest, gallery forest, mangroves. N Colombia, in R. César Valley, W into Sierra Nevada de Santa Marta in E Magdalena Department.

ID Palest of N Colombian and Venezuelan gracile capuchins; coat colour variable; cap cinnamon to brown; face, throat, neck cartridge buff; back, forearms, forelegs orangey; tail frosted cinnamon-brown.

△ Up to 500 m

Santa Marta White-fronted Capuchin *Cebus malitiosus* EN
ES **Mono cariblanco**

CO **E**; N Andes.

H Lowland, submontane, and montane forest. N Colombia, known only from the NW base of Sierra de Santa Marta.

ID Cap pale brown; back cinnamon brown; forearms and forelegs not markedly contrasting in colour with back and sides of body; belly and chest ochraceous tawny to cinnamon-brown and silvery; contrasting pale area of front extends well over upper surfaces of shoulders and inner sides of upper arms.

HB c. 45.7 cm
T c. 43.3 cm

Trinidad White-fronted Capuchin *Cebus trinitatis* CR
EN **Weeping Capuchin, Matchin**

TT **E**; Caribbean island.

H On E & SE of Trinidad I, lowland moist forests of the Nariva Swamp, lowland forest E of the Central Range Mts., and the Trinity Hills Mts.

No measurements available

ID Almost completely blond; light brown cap; pink face; greyish hands and feet; sometimes orangey forelimbs.

HB 35–51 cm
T 40–50 cm
W 1.7–3.6 kg (M), 1.2–2.2 kg (F)

Ecuadorian White-fronted Capuchin
Cebus aequatorialis CR

ES **Capuchino ecuatorial, Machín blanco**

EC, PE; Pacific coast, W Andes.

H Dry forest in lowlands and including coastal areas to wet submontane Andean forest. Ecuador lowlands W of the Andes and NW Peru.

ID Head pale yellowish-white; brown cap; black line to nose; upperparts pale cinnamon rufous; darker along midline of back; hands and feet more brownish than arms and legs; ventral surface paler than flanks.

△ 1100–2040 m

Colombian White-faced Capuchin *Cebus capucinus capucinus* EN
ES **Maicero capuchino**, **Mono capuchino**, **Capuchino de cara blanca**

CO, EC, PA; C America.

H *Terra firme* forest, seasonally inundated forests, palm forests, mangroves, deciduous dry forest. On the W slopes of the Andes, E Panama, W Colombia, and NW Ecuador.

ID Black cap with white or yellowish-white on face, front of crown, throat, and shoulders; body, crown, limbs, and tail black; chest, shoulders and upper arms white.

△ Up to 1800–2100 m

HB 40–49 cm
T 45–55 cm
W 3–4.3 kg (M),
1.8–3.5 kg (F)

Gorgona White-faced Capuchin
Cebus capucinus curtus VU
ES **Mono capuchino**

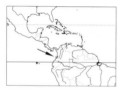

CO E; Pacific Is.

H Humid submontane forest of Gorgona I, Colombia.

ID Black cap and body; chest white, extending forward to face and front of crown and upward to shoulders and upper arms.

HB 32–33 cm
T 42 cm
W 2.5–3 kg (M),
1.6–3 kg (F)
(smaller than *C. c. capucinus*)

Panamanian White-faced Capuchin *Cebus imitator* VU
ES **Mono carablanca**, **Mono capuchino**

CR, HN, NI, PA; C America.

H Lowland, submontane, and montane moist forest, tropical dry forest, gallery forest, mangroves. N Honduras, C & W Nicaragua, Costa Rica, and W Panama (including Coiba and Jicarón Is.).

ID Like *C. c. capucinus*, but some females have elongated frontal tufts with brownish tinge.

△ Up to 1800–2100 m

HB 34.3–42 cm (M),
38.5–40.5 cm (F)
T 44–46 cm (M),
43–45 cm (F)
W 3.7–3.9 kg (M),
2.6–2.7 kg (F)

NEOTROPICAL PRIMATES

NIGHT MONKEYS · Aotidae

Panamanian Night Monkey *Aotus zonalis* NT
ES **Mono nocturno, Marteja**

HB c. 30 cm
T c. 36 cm
W 890–916 g

CO, PA; C America.

H Lowland forest in the Pacific lowlands of Panama, but absent from SW Panama (Chiriquí), NW Colombia, N into Córdoba Department, S to the R. Raposo, E to the Sinú Valley and possibly N Antioquia.

ID Body brownish in the W Canal Zone, grading into paler and greyer tones in E Panama; hair, short, appressed. In Colombia, similar to *A. griseimembra* except for its dark brown or blackish hands and feet.

△ Up to 650 m

Lemurine Night Monkey *Aotus lemurinus* VU
ES **Mono nocturno lemurino, Marteja**

HB c. 31 cm
T c. 36 cm
W 800–1050 g

CO, EC, VE; N Andes.

H Tropical high-elevation and humid subtropical forest.

ID Grey-necked; upperparts variable, greyish/buffy-agouti; poorly defined brownish medial dorsal band; underparts yellowish to pale orange; limbs, hands, feet greyish-agouti; facial pattern variable; temporal stripes separate or united behind head; no interscapular whorl or crest; underside of tail red, tip blackish red.

△ 1000–3200 m

Grey-legged Night Monkey *Aotus griseimembra* VU
ES **Mono nocturno, Marteja, Mico de noche caribeño**

CO, VE; N Andes.

H Primary and secondary lowland forest in the R. Magdalena Valley in Colombia.

HB c. 34 cm
T c. 34 cm
W 800–1000 g

ID Grey-necked; side of neck greyish to brownish agouti; upperparts greyish to buffy; poorly defined brownish medial dorsal band; chest and underparts brownish or yellowish to pale orange; pelage short, appressed; facial pattern inconspicuous; hands and feet pale coffee-brown, darker hair tips; no interscapular whorl or crest.

Brumback's Night Monkey *Aotus brumbacki* VU
ES **Mono nocturno**, **Marteja**, **Mico de noche llanero**

CO **E**; NE Andes, Llanos.
H Lowland closed canopy forest, gallery forest, N of the R. Guayabero to the R. Arauca, C Colombia.
ID Grey-necked; back greyish with dark brown mid-dorsal zone; underparts pale orange to elbows, knees, throat; side of neck, flank, outer side of arms greyish or brownish agouti; thin, brownish-black temporal stripes; yellowish above eye; white on face extends to chin; short, longitudinal interscapular crest.
△ Up to 1500 m

No measurements available

Humboldt's Night Monkey *Aotus trivirgatus* LC
PT **Macaco-da-noite**, ES **Mono nocturno**, **Marteja**

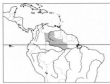

BR, CO, VE; N Amazonia.
H Lowland, submontane and dry forests, E of the R. Negro, W of the R. Trombetas in Brazil, SC Venezuela (Orinoco and W of the R. Caura) into E Colombia.
ID Grey-necked; side of neck greyish to brownish; upperparts greyish to buffy; chest, belly, inside limbs orange-buffy; face greyish; triradiate stripe brown; hands and feet dark brown; parallel temporal stripes; no interscapular whorl or crest.
△ 0–1000 m

HB 30–38 cm
T 33–40 cm
W c. 775 g

Spix's Night Monkey *Aotus vociferans* LC
PT **Macaco-da-noite**, ES **Mono nocturno de Spix**, **Marteja**, **Musmuqui**

BR, CO, EC, PE; W Amazonia.
H Lowland, *terra firme*, seasonally-flooded and swamp forest, W of the R. Negro.
ID Grey neck; upperparts and crown brownish-greyish; underparts pale orange; white chin and spots above eyes; hands and feet black; dark temporal and mid-crown stripes; malar stripes distinct to absent; interscapular whorl; ventral side of tail reddish or greyish-red, black towards tip.
△ 200–900 m

HB 35–45 cm

T 31–47 cm
W 700–800 g

Hernández-Camacho's Night Monkey *Aotus jorgehernandezi* DD
ES **Mono nocturno, Marteja**

CO **E**; N Andes.

H Submontane and possibly montane forest on the W slopes and foothills of the Andes of W Colombia in the region of Quindío and Risaralda.

No measurements available

ID Grey-necked; white spots above eyes, separated by black frontal stripe; subocular white bands of fur separated by thin black malar stripe on each side of head; chest, belly, and inside limbs with thick white fur.

Andean Night Monkey *Aotus miconax* EN
ES **Musmuqui, Mono lechuza, Tutamono, Mono nocturno andino**

PE **E**; N Andes.

H Pre-montane and montane cloud forest, S of the R. Marañón and W of the R. Huallaga to about 10° S.

HB c. 39.4 cm
T c. 22 cm
(from type specimen)

Red-necked; dorsum brownish light grey; ventrum to chin and inner sides of limbs pale orange; white around eyes, sides of nose, and chin; two black temporal stripes and one diamond-shaped on mid-crown fanning out from forehead; tail blackish above, orangey below; no intrascapular whorl.

△ 1500–3100 m

Ma's Night Monkey *Aotus nancymai* VU
PT **Macaco-da-noite**, ES **Mono nocturno, Marteja, Mono lechuza**

HB 29–34 cm
T 35–42 cm
W 780–795 g

BR, CO, PE; W Amazonia.

H Lowland *terra firme*, seasonally flooded and swamp forest, S of the Solimões/Amazonas, W possibly from the R. Jandiatuba to the R. Huallaga in Peru.

ID Red-necked; dorsum greyish-agouti; inner sides of limbs, chest, ventrum pale orange; face greyish-white; malar, mid-crown, and narrow temporal stripes blackish; silvery buff chin and patches below eyes; proximal third of tail orange, the rest black.

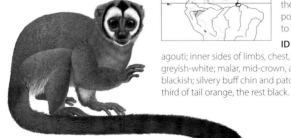

HB 35–42 cm
T 35–44 cm
W c. 875 g (M),
c. 1040 g (F)

Black-headed Night Monkey *Aotus nigriceps* LC
PT **Macaco-da-noite**, ES **Cuatro ojos, Mono-michi, Musmuqui**

BO, BR, PE; SW Amazonia.

H *Terra firme*, seasonally-flooded and swamp forest, S of the R. Amazonas/Solimões, W of the R. Tapajós–Juruena to the rios Guaporé and Madre de Dios, and R. Huallaga.

ID Red-necked; body iron-grey; ventrum, throat, chin, inside of limbs, and proximal third of tail orange; crown, broad face stripes; tail black; white on face and chin conspicuous; no interscapular whorl.

HB 33–34 cm
T c. 31 cm
W 990–1580 g

Azara's Night Monkey *Aotus azarae azarae* DD
PT **Macaco-da-noite**, ES **Miriquiná, Mono nocturno, Lechuza**

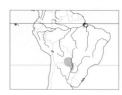

AR, BO, BR, PY; Chaco, Pantanal.

H Deciduous, gallery and secondary forest of the W and N margins of the Pantanal, Cerrados del Chaco and Dry Chaco, W of the R. Paraná.

ID Red-necked; long, shaggy fur, greyish to pale buffy above, pale whitish-orange below; temporal stripes narrow; median stripe lozenge-shaped; face white above and around eyes; chin whitish; tail reddish, black tip; interscapular whorl usually present.

HB 33–34 cm
T c. 31 cm
W 990–1580 g

Bolivian Night Monkey *Aotus azarae boliviensis* DD
ES **Cuatro ojos, Mono nocturno, Lechuza**

BO, PE; E Andes, Chiquitano, Chaco.

H *Terra firme* and gallery forest, Beni savannah, Chiquitano dry forest, and dry Chaco. S of the ríos Madre de Dios and Inambari, to the Bañados del Izozog, W of the R. Paraguay.

ID Red-necked; fur short, with olive tone above and greyer on limbs; narrow black face stripes; medial stripe expands on crown; tail orange, black tip; interscapular whorl.

HB 33–34 cm
T c. 31 cm
W 990–1580 g
(as for nominate ssp.)

Feline Night Monkey
Aotus azarae infulatus LC
PT **Macaco-da-noite**

BR E;
E Amazonia.

H Humid, deciduous and palm forests, Cerrado and Pantanal, S of the R. Amazonas (enclave in SE Amapá), between the rios Tapajós–Juruena and Parnaíba–Tocantins.

ID Red-necked; body silvery grey; ventrum, inside limbs, throat pale orange; face white; chestnut to black malar and lateral stripes; median stripe forms rhomboid on crown; interscapular whorl usually present.

TITI MONKEYS, SAKIS, BEARDED SAKIS AND UACARIS · Pitheciidae

Río Beni Titi *Plecturocebus modestus* EN
ES Lucachi

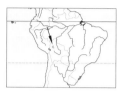

BO E ; N Bolivia.

H Dry forest in a forest-savannah mosaic, E of the R. Beni and W to the R. Manique.

ID Body mainly light brownish or reddish agouti; forehead and crown reddish-brown agouti; pale, whitish ear tufts; thin, blackish fringe on forehead; sideburns same colour as forehead and crown; hands and feet blackish mixed with red; tail blackish-agouti, and notably greyer than body.

△ Up to 400 m

HB c. 31.5 cm (M)
T c. 40 cm (M)
W c. 800 g

HB c. 32.5 cm (M)
T c. 42.5 cm (M)
W c. 800 g

Olalla Brothers' Titi
Plecturocebus olallae CR
ES Ururo

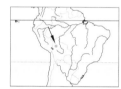

BO E ; N Bolivia.

H Gallery forest patches in a forest-savannah mosaic in N Bolivia, along the ríos Yacuma and Manique.

ID Body fur fluffy, uniformly orange; outer surfaces of limbs reddish-brown; sideburns, beard, and forehead blackish, forming a facial fringe; crown reddish-brown agouti; small, whitish ear tufts; hands and feet black; tail dark agouti, contrasting with orange back.

△ Up to 400 m

HB 28–42 cm
T 37–46 cm
W c. 800 g

White-eared Titi
Plecturocebus donacophilus LC
PT Zogue-zogue-de-orelha-branca, ES Faca-faca, Luchachi

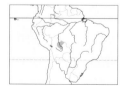

BO, BR; S Amazonia, Pampas.

H Rainforest and dry forest patches in the ecotone with the Bolivian pampas. E of the Río Manique, in forest patches woodland savannah, E to the upper R. Jiparaná.

ID Body mainly buff or greyish-agouti to orange agouti; chest, belly, and inside limbs orange; hands and feet buffy or buffy agouti, lighter than rest of body; distinct whitish ear tufts and malar stripe; tail buffy to whitish.

NEOTROPICAL PRIMATES

Pale Titi *Plecturocebus pallescens* LC
PT **Zogue-zogue-pálido**, ES **Ururu, Mono tití, Tití chaqueño**

BO, BR, PY; Chaco.

H Humid forest, riparian and swampy forest in the Cerrados del Chaco, Bolivia, continuous with the dry forest of the Paraguayan northern Chaco, W of the R. Paraguay, and the Pantanal wetlands in Brazil.

ID Head and body with long, shaggy hair, pale buff agouti; forehead, sideburns, ear tufts whitish to white; evident dark malar stripe; hairs almost conceal facial skin; limbs and tail whitish.

HB 31–36 cm
T 39–42 cm
W likely c. 800 g

HB 30–36 cm
T 38–45 cm
W 1160–1180 g

Ornate Titi *Plecturocebus ornatus* VU
ES **Zogui zogui**

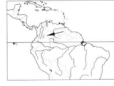

CO **E**; NE Andes, Llanos.

H Lowland *terra firme* and secondary forest in the Cundinamarca Department, Colombia, N to the lower ríos Upía and Meta, and S through the Sierra de la Macarena to the ríos Guayabero and Guaviare.

ID Body and crown buff-agouti; underparts, inside limbs, forearms, lower legs, sideburns, throat reddish; forehead, ear tufts, hands, feet whitish; tail whitish, dark red-brown at base.

HB c. 35 cm
T c. 61 cm
(from type specimen)
W c. 800 g

Caquetá Titi *Plecturocebus caquetensis* CR
ES **Zogui zogui**

CO **E**; NW Amazonia.

H Degraded forest patches in predominantly pasture and agricultural land, N of the upper R. Caquetá, N to its tributary, the R. Orteguaza, in S Colombia.

ID Dorsum pale brown; neck, flanks, upper arms, and legs more greyish-brown; crown light buffy brown; underparts, forearms, lower legs, beard, and sideburns orangey-red; tail greyer than brown, paler towards tip.
△ 190–400 m

dark form

HB 30–32 cm
T 36–40 cm
W likely c. 800 g

San Martín Titi
Plecturocebus oenanthe **CR**

ES **Mono tocón de San Martín**

PE **E**;
E Andes.

H Montane cloud forest, inundated areas, *Mauritia* palm swamp, *Ficus*-dominated forest, low secondary forest and bamboo stands, W of the R. Huallaga, in the R. Mayo Valley, S to the R. Huayabamba, Peru.

ID Body dark brown agouti or light brown agouti; underparts and inside limbs orange; forehead blaze and long circumfacial crest buffy or whitish; pale malar stripe; tail dark brown.

△ 750–950 m

HB 27–41 cm
T 39.5–48 cm
W 1000–1200 g

Coppery Titi *Plecturocebus cupreus* **LC**

PT **Zogue-zogue-acobreado**, ES **Tocón colorado**

BR, PE; W Amazonia.

H Lowland *terra firme* forest in the understorey of the middle and lower strata, S of the rios Marañón–Solimões, E of the R. Ucayali, in E Peru, and W of the R. Purus, Brazil.

ID Body buffy-brown agouti; underparts, forearms, lower legs, throat, sideburns, hands, feet reddish; crown and back reddish-brown; forehead blackish; tail greyish-white; tails of young brown until 2 years old.

△ 100–300 m

NEOTROPICAL PRIMATES

Red-crowned Titi *Plecturocebus discolor* LC
ES **Cotoncillo rojo, Tití rojizo, Zogui zogui**

CO, EC, PE; W Amazonia.

H Lowland rain forest in understorey and lower strata, S of the R. Guamués, S of the R. Napo–Aguarico, E of the Central Cordillera Cahuapanas across the R. Huallaga, N of the R. Mayo, to the R. Ucayali.

HB 28–35 cm
T 38–51 cm
W 850–1080 g

ID Body agouti; underparts, forearms, lower legs, crown, hands, feet, sideburns, throat reddish; forehead white; ear tufts whitish; tail half brownish-grey, half pale cream-colour.

△ 100–980 m

HB 30–41 cm
T 38–47.6 cm
W c. 800 g

Chestnut-bellied Titi
Plecturocebus caligatus LC
PT **Zogue-zogue**

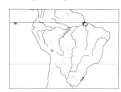

BR **E**; C Amazonia.

H Lowland *terra firme* rain forest in the denser understorey of the middle and lower strata, S of the R. Solimões/Amazonas, between the rios Purus and Madeira, S to the R. Ipixuna.

ID Body predominantly reddish-brown agouti; underparts, inside limbs, throat, sideburns, and forearms reddish, chestnut; hands and feet blackish to reddish-brown; forehead and anterior portion of crown black.

△ 100–200 m

HB 37–40 cm
T 39–47 cm
W c. 800 g

Doubtful Titi *Plecturocebus dubius* LC
PT **Zogue-zogue-de-Hershkovitz**

BR **E**; W Amazonia.

H Lowland forest in dense understorey, between the middle rios Purus and Madeira, S from the R. Mucuím to the rios Madeira–Abunã.

ID Dorsum and crown brownish-agouti; underparts, inside limbs, forearms, lower legs, sideburns, and beard reddish; hands and feet blackish; forehead whitish, bordered below by blackish line; tail proximal one-third like dorsum, distal two-thirds blackish.

△ 100–400 m

Stephen Nash's Titi *Plecturocebus stephennashi* DD
PT **Zogue-zogue-de-Stephen-Nash**

BR E ; W Amazonia.

H Lowland forest, understorey of the middle to lower strata, in the Ipixuna–Mucuím interfluvium on the right bank of the R. Purus. Possibly a hybrid of *P. caligatus* × *P. dubius*.

ID Dorsum, crown, rump silvery or mixed buffy brown; underparts and inside limbs dark red; hands and feet silvery buff; forehead and anterior crown black; proximal tail silvery, mixed with brownish, distal white or buffy.

△ 100–200 m

HB 27–28 cm
T c. 42 cm
W c. 780 g

HB 32–40 cm
T 39–48 cm
W 740–950 g

Ashy Titi *Plecturocebus cinerascens* LC
PT **Zogue-zogue-cinzento**

BR E ; C Amazonia.

H Lowland *terra firme* forest, white-sand forest (*campinarana*), and riparian forests in the denser understorey. S of the R. Madeira, S between the rios Canumã–Acari and Guariba, east to the upper rios Sucunduri and middle Juruena into Rondônia.

ID Body grey to dark grey; tawny agouti mid-dorsal region; sideburns and throat greyish to yellowish-agouti; tail blackish to greyish.

HB 26–35 cm
T 39.5–48 cm
W 1040–1500 g

Milton's Titi *Plecturocebus miltoni* DD
PT **Zogue-zogue-de-rabo-de-fogo**

BR E ; W Amazonia.

H Lowland *terra firme* forest in the denser understorey, between the rios Roosevelt and Guariba, S to about 12° 30′ S.

ID Dorsum, flanks, legs, arms, and crown grey; underparts and inside limbs orange; throat, beard, and sideburns deep orangey-brown; whitish stripe on front of crown; whitish patches above eyes; hands and feet grey to whitish; tail orange.

NEOTROPICAL PRIMATES

Hoffmanns's Titi *Plecturocebus hoffmannsi* LC
PT **Zogue-zogue-de-Hoffmanns**

BR **E** ; E Amazonia.

H Lowland *terra firme* forest and swamp forest in the denser understorey, S of the R. Amazonas, between the R. Canumã and Urariá, and Ramos channels ('paranás') and the R. Tapajós.

ID Dorsum chestnut-brown; underparts, inside limbs, hands, feet, cheeks, and chin pale orange-yellowish; crown grey; face black; outer limbs greyish-brown; hands and feet blackish; tail blackish.

HB 27–36 cm
T 40–53 cm
W 920–1100 g

HB 30–41
T 42–50 cm
W 930–1400 g

Lake Baptista Titi *Plecturocebus baptista* LC
PT **Zogue-zogue-do-Lago-Baptista**

BR **E** ; E Amazonia.

H Lowland *terra firme* and swamp forest in the denser understorey, S of the rios Amazonas and lower Madeira, S the Urariá and Ramos channels ('paranás') and between the rios Uira–Curapá and Mamurú.

ID Dorsum dark brown; underparts, inside limbs, forearms, lower legs bright reddish or reddish-brown; crown greyish; face black; tail black, sometimes with white tip.

HB 27–43 cm
T 35–55 cm
W 700–1220 g

Red-bellied Titi *Plecturocebus moloch* LC
PT **Zogue-zogue-vermelho-inchado**

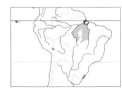

BR **E** ; E Amazonia.

H Understorey of lowland forest, S of the R. Amazonas between the rios Tapajós and Tocantins, excepting the ranges of *P. vieirai*, between the rios Xingu and Iriri, and *P. grovesi*, W of the R. Teles Pires.

ID Dorsum flanks and outside limbs pale-brown agouti; underparts, inside limbs, throat, sideburns light orange; face black; hands and feet pale; tail pale, with blackish to orange banding.
△ Up to 300 m

Vieira's Titi *Plecturocebus vieirai* CR
PT **Zogue-zogue-de-Vieira**

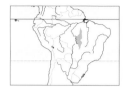

BR **E**; E Amazonia.

H Lowland rainforest, riparian forest, and *Mauritia flexuosa* palm swamp, S of the R. Amazonas, between the rios Xingu and Iriri in the states of Pará and Mato Grosso.

ID Dorsum, crown, nape, outer limbs pale greyish to reddish-brown; underparts and inside limbs pale orangey; hands and feet white; forehead, cheeks, and beard whitish; black face; tail slightly darker than back, tip paler.

HB 30–35 cm
T 41–51 cm
W c. 960 g

HB 32–34 cm
T 43–44 cm
W c. 960 g

Groves's Titi *Plecturocebus grovesi* CR
PT **Zogue-zogue-do-Mato-Grosso**

BR **E**; E Amazonia.

H Lowland *terra firme* forest in the denser understorey, from the rios Juruena and Arinos in the W to the R. Teles Pires in the E, S to about 10° S, in the state of Mato Grosso.

ID Dorsum pale greyish-brown; flanks, outside limbs, crown pale greyish; underparts, inside limbs, cheeks, beard rusty red; face black; forehead a little paler than crown; hands and feet whitish; tail black, white tip.
△ 100–200 m

HB 30–34.5 cm
T 38–44 cm
W c. 850 g

Brown Titi *Plecturocebus brunneus* VU
PT **Zogue-zogue-marrom**

BR **E**; S Amazonia.

H Lowland *terra firme* rain forest in the understorey of the middle and lower strata, E of the Madeira, between the rios Mamoré and Jiparaná, S along the R. Guaporé to about 13° S in Rondônia.

ID Dorsum, flanks, arms, legs, and crown greyish-brown; underparts brownish or reddish; forehead blackish; sideburns and beard dark reddish-brown; no ear tufts; tail brownish-grey, whitish towards tip.
△ 100–400 m

Prince Bernhard's Titi *Plecturocebus bernhardi* LC
PT **Zogue-zogue-do-Principe-Bernhard**

BR **E**; C Amazonia.

H Lowland forest, secondary and disturbed forest, and rubber plantations. S of the R. Madeira, W of the rios Aripuanã and Roosevelt to the R. Jiparaná. Rondônia.

ID Dorsum, flanks, legs, and arms greyish-black, mixed with brownish or rust-brown on back; underparts, inside limbs, cheeks, beard dark orange; face black; whitish ear tufts; hands and feet white; tail black with white tip.

HB 36–38 cm
T c. 55 cm
W 700–1200 g

No measurements available

Toppin's Titi *Plecturocebus toppini* LC
PT **Zogue-zogue-de-Toppin**, ES **Tocón colorado**

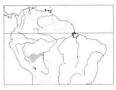

BO, BR, PE; SW Amazonia.

H Lowland *terra firme* rain forest in the understorey. S of the R. Purus, W from the R. Ituxi, S to the R. Madre de Dios, W to the ríos Ucayali and Manu, Peru.

ID Dorsum, flanks, upper arms and legs, and crown grizzled brown, the last more yellowish; underparts, lower legs, forearms, hands, feet, sideburns, beard, and small ear tufts red-brown; forehead black; tail blackish, paler towards tip.

HB c. 30 cm
T c. 40 cm
W c. 900 g

Urubamba Brown Titi *Plecturocebus urubambensis* LC
ES **Tocón de Urubamba**

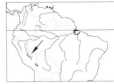

PE **E**; SW Amazonia.

H Understorey of lowland forest, S of the ríos Inuya and Tambo, W of the R. Urubamba (to confluence with the R. Camisea) and R. Manu to the upper Madre de Dios, SE Peru.

ID Dorsum, flanks, upper arms, legs, underparts, and crown brownish-agouti; forehead, sideburns, and malar stripe black mixed with agouti; hands, feet, forearms black; base of tail blackish, mixed with agouti, pale near end.

Madidi Titi *Plecturocebus aureipalatii* LC
ES **Lucachi, Luca-luca**

BO, PE; E Amazonia.

H Understorey of lowland *terra firme* rain forest, E from the Andean foothills in Bolivia, between the ríos Madre de Dios and Beni, into SE Peru.

ID Dorsum, flanks, upper legs and arms light brown; underparts pale orange, sparsely haired; chest, throat, sideburns, forearms, and lower legs orangey-red; face black; crown brownish with golden sheen; tail dark at base, fading to white tip.

△ 100–500 m

HB 29–32 cm
T 52–48 cm
W 900–1000 g

HB 33–36 cm
T 42.5–49 cm
W 1100–1460 g

Medem's Titi
Cheracebus medemi VU
ES **Zogui zogui, Viudita**

CO **E**; NW Amazonia.

H Lowland *terra firme* forest and white-sand forest in SW Colombia, E from the Andean foothills between the upper ríos Caquetá and Putumayo. The E limit of its distribution is not known.

ID Head, body, tail, and underparts entirely or predominantly black or blackish; throat and chest white; face black, some white on cheeks and chin; hands blackish.

△ 100–450 m

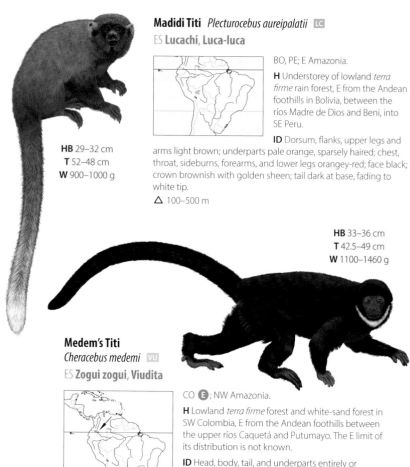

HB 36–46 cm
T 37–50 cm
W 1000–1500 g

White-collared Titi *Cheracebus torquatus* LC
PT **Zogue-zogue-de-colarinho-branco**

BR **E**; C Amazonia.

H *Terra firme* and black-water seasonally inundated forest, S of the R. Solimões between the rios Juruá and Purus, S to the R. Tapauá.

ID Dorsum, flanks, legs dull brown, reddish-brown to reddish; underparts and legs rusty reddish; crown bright red; white whiskers; collar distinct, yellowish to whitish, extending to ears; hands yellowish; forearms and feet blackish; tail black, mixed with reddish.

△ 100–200 m

White-chested Titi *Cheracebus lugens* LC
PT **Zogue-zogue-de-peito-branco**, ES **Zogui zogui, Mono viudita**

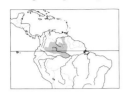

BR, CO, VE; N & W Amazonia.

H Lowland and montane, white-sand and swamp forest, S of the Orinoco, between the ríos Caura and Caroní, S of the R. Ventuari, W from the rios Branco and Negro, N of the R. Japurá/Caquetá, N to the R. Tomo in Colombia.

ID Dorsum, flanks, crown dark red to brown; underparts dark brown/blackish; forehead, forearms, feet, and tail blackish; collar white; hands yellowish.

△ 100–1000 m

HB 30–40 cm
T 41–49 cm
W 1000–1500 g

Yellow-handed Titi *Cheracebus lucifer* LC
PT **Zogue-zogue-de-mão-amarela**, ES **Tití de manos amarillos, Tocón negro**

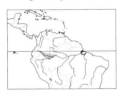

BR, CO, EC, PE; W Amazonia.

H Lowland, white-sand or sandy-clayey soil forests, S of the R. Japurá/Caquetá and N of the rios Solimões/Amazonas in Peru and R. Aguarico.

ID Body entirely blackish, with brownish/reddish on back and flanks; underparts, forehead, sideburns, feet, and tail blackish; crown reddish; cheeks and chin with white whiskers; hands orangey; collar white, extending to ears.

△ 100–300 m

Juruá Collared Titi *Cheracebus regulus* LC
PT **Zogue-zogue-do-rio-Juruá**

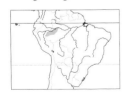

BR **E**; W Amazonia.

H Lowland *terra firme* rainforest, S of the R. Solimões, between the lower R. Javarí in the W, and lower R. Juruá in the E, S to about 7° S, but not really known.

HB 35–41 cm
T 45–48 cm
W c. 1500 g

ID Body, flanks, legs, and feet brown or blackish; inside arms blackish; distinct reddish crown; hands orange; collar white, extending to ears; sideburns brownish; tail blackish.

△ 100–200 m

HB 37–45 cm
T 44–49 cm
W 1000–1500 g

Aquino's Collared Titi *Cheracebus aquinoi* NE
ES **Mono tocón de Aquino**

PE **E**; W Amazonia.

H Sandy soil forest with thin-stemmed trees, gallery forests, flooded forests, and palm-dominant forests between the ríos Nanay and Tigre, W bank affluents of the lower R. Ucayali.

ID Overall dark reddish-brown; limbs brownish; black band across forehead; face with sparse off-white hairs; collar restricted to cream-coloured bow-tie; creamy-white hands; contrasting blackish tail.

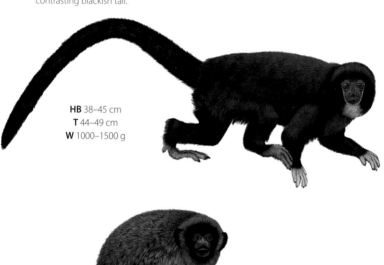

HB 38–45 cm
T 44–49 cm
W 1000–1500 g

HB 30–39.5 cm
T 45–50 cm (F)
W 1000–1600 g

Black-fronted Titi *Callicebus nigrifrons* NT
PT **Sauá-de-cara-preta**

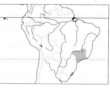

BR **E**; Atlantic Forest, Cerrado.

H Primary and disturbed forest, forest patches in Cerrado. From the rios Paraná and Parnaíba, E to the Mantiqueira and Espinhaço ranges, N to the upper R. São Francisco, S to the R. Tietê.

ID Body brownish to orange-brown; underparts, throat, sideburns, and crown pale brownish; forehead, ears, cheeks, chin blackish; hands and feet black; tail orange to rusty red-brown.

△ Up to 1000 m

NEOTROPICAL PRIMATES

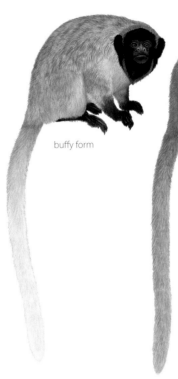

buffy form

orange form

HB 31–42 cm
T 42–56 cm
W 1000–1650 g

Masked Titi *Callicebus personatus* VU
PT **Guigó-mascarado, Sauá**

BR E ; Atlantic Forest.

H Primary and disturbed forest, N from Rio de Janeiro state, through Espírito Santo to the R. Mucuri and, inland, the R. Jequitinhonha, W to the Serra da Mantiqueira, Minas Gerais state.

ID Dorsum, flanks, limbs golden-orange in N of range, buffy brown in S; hands and feet black; black throat, sideburns, forehead, face, and crown form striking mask.

△ Up to 1000 m

HB 33–37 cm
T 39.5–51 cm
W c. 1400 g

Southern Bahian Titi *Callicebus melanochir* VU
PT **Guigó-do-sul-da-Bahia**

BR E ; Atlantic Forest.

H Primary and disturbed coastal Atlantic Forest in S Bahia and N Espírito Santo states, from the R. Paraguaçú in the N to the R. Mucurí in the S. Limited to the W by the transition to semi-arid caatinga scrubland.

ID Body and tail greyish-agouti or buffy brownish-agouti; hands, feet, facial fringe outlining face, and ears blackish; back and base of tail washed with reddish-brown.

△ Up to 300 m

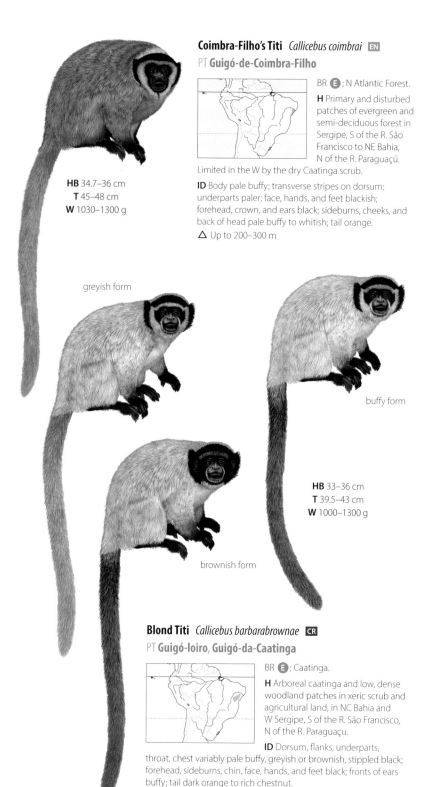

Coimbra-Filho's Titi *Callicebus coimbrai* EN
PT **Guigó-de-Coimbra-Filho**

BR **E** ; N Atlantic Forest.

H Primary and disturbed patches of evergreen and semi-deciduous forest in Sergipe, S of the R. São Francisco to NE Bahia, N of the R. Paraguaçú. Limited in the W by the dry Caatinga scrub.

ID Body pale buffy; transverse stripes on dorsum; underparts paler; face, hands, and feet blackish; forehead, crown, and ears black; sideburns, cheeks, and back of head pale buffy to whitish; tail orange.

△ Up to 200–300 m

HB 34.7–36 cm
T 45–48 cm
W 1030–1300 g

greyish form

buffy form

brownish form

HB 33–36 cm
T 39.5–43 cm
W 1000–1300 g

Blond Titi *Callicebus barbarabrownae* CR
PT **Guigó-loiro, Guigó-da-Caatinga**

BR **E** ; Caatinga.

H Arboreal caatinga and low, dense woodland patches in xeric scrub and agricultural land, in NC Bahia and W Sergipe, S of the R. São Francisco, N of the R. Paraguaçu.

ID Dorsum, flanks, underparts, throat, chest variably pale buffy, greyish or brownish, stippled black; forehead, sideburns, chin, face, hands, and feet black; fronts of ears buffy; tail dark orange to rich chestnut.

△ 240–900 m

NEOTROPICAL PRIMATES

HB 32–41.5 cm
T 37–45.5 cm
W c. 1.5 kg

White-faced Saki *Pithecia pithecia* LC
PT **Parauacu-de-cara-branca**, ES **Mono viudo**, FR **Saki à face pâle**, SR **Wanakoe**, GY **White-faced huruwa**

BR, GF, GY, SR; E & N Amazonia.

H *Terra firme* forest, *Mauritia* palm swamp, seasonally flooded forest and secondary growth. N of the R. Amazonas, E of the rios Negro–Branco to the Orinoco, E of the lower R. Caroni, Venezuela, and E of the R. Trombetas, Brazil.

ID Female greyish mottled pelage; bright orange chest. Male black with white facial disk. Juvenile/sub adult male resembles female.

HB 28.5–46 cm
T 34–45 cm
W c. 1.9 kg

Golden-faced Saki *Pithecia chrysocephala* LC
PT **Parauacu**

BR **E** ; C Amazonia.

H *Terra firme* forest, *Mauritia* palm swamp, seasonally flooded forest and secondary growth, N of the R. Amazonas, both sides of the lower R. Negro, E to Faro and the R. Nhamundá.

ID Female greyish, mottled pelage; bright orange chest; black facial hair, orange malar lines; white star on forehead. Male black; orange-reddish facial disk. Juvenile/sub adult male resembles female.

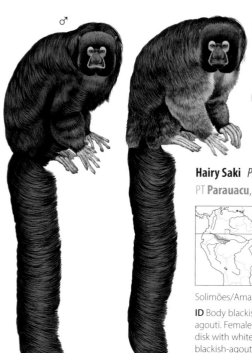

Hairy Saki *Pithecia hirsuta* LC
PT **Parauacu**, ES **Mono volador, Huapo negro**

BR, CO, PE; W Amazonia.

H Lowland *terra firme* forest, palm swamp and seasonally inundated forest between the rios Solimões/Amazonas and Napo and Japurá/Caquetá.

ID Body blackish, stippled white; crown brownish-agouti. Female more extensive stippling; black facial disk with white malar stripe but no moustache. Male blackish-agouti to brown head; black chest ruff; white malar stripe and moustache.

Miller's Saki *Pithecia milleri* VU
ES **Parahuaco, Saki de Miller, Mono volador, Huapo**

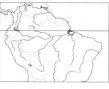

CO, EC, PE; W Amazonia.

H Lowland *terra firme* forest, palm swamp, between the upper ríos Napo and Caquetá–Caguán, NE to the Sierra de la Macarena.

ID More grizzled greyish than *P. hirsuta*. Female more greyish; crown pale greyish; whitish crown malar stripes; indistinct moustache. Male darker; underparts, crown, ruff, wrists, lower legs brown; whitish malar stripe and moustache; tail blackish with white hair tips.

△ Up to 700 m

NEOTROPICAL PRIMATES

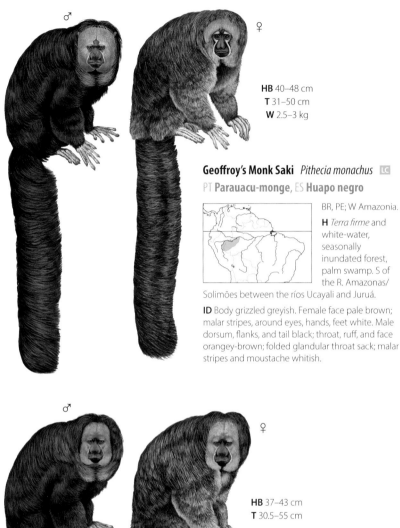

HB 40–48 cm
T 31–50 cm
W 2.5–3 kg

Geoffroy's Monk Saki *Pithecia monachus* LC
PT **Parauacu-monge**, ES **Huapo negro**

BR, PE; W Amazonia.

H *Terra firme* and white-water, seasonally inundated forest, palm swamp. S of the R. Amazonas/Solimões between the ríos Ucayali and Juruá.

ID Body grizzled greyish. Female face pale brown; malar stripes, around eyes, hands, feet white. Male dorsum, flanks, and tail black; throat, ruff, and face orangey-brown; folded glandular throat sack; malar stripes and moustache whitish.

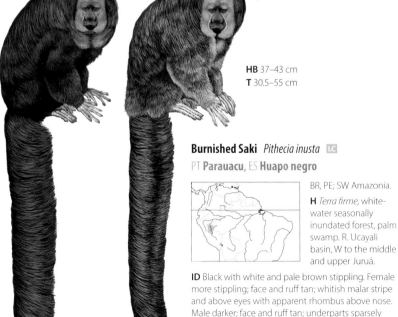

HB 37–43 cm
T 30.5–55 cm

Burnished Saki *Pithecia inusta* LC
PT **Parauacu**, ES **Huapo negro**

BR, PE; SW Amazonia.

H *Terra firme*, white-water seasonally inundated forest, palm swamp. R. Ucayali basin, W to the middle and upper Juruá.

ID Black with white and pale brown stippling. Female more stippling; face and ruff tan; whitish malar stripe and above eyes with apparent rhombus above nose. Male darker; face and ruff tan; underparts sparsely haired black; wrists black; hands and feet whitish; indistinct malar stripe.

101

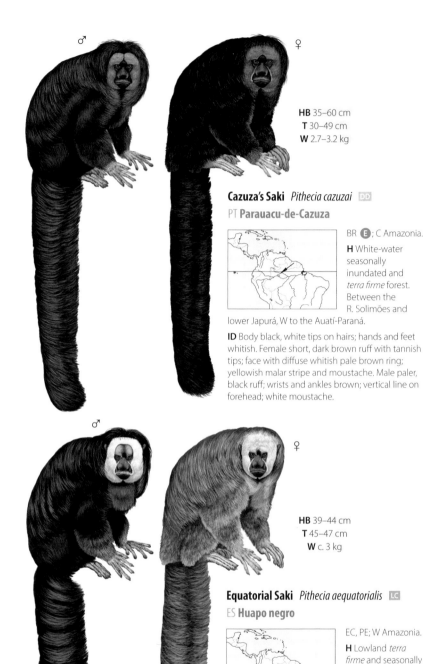

HB 35–60 cm
T 30–49 cm
W 2.7–3.2 kg

Cazuza's Saki *Pithecia cazuzai* DD
PT **Parauacu-de-Cazuza**

BR **E**; C Amazonia.
H White-water seasonally inundated and *terra firme* forest. Between the R. Solimões and lower Japurá, W to the Auatí-Paraná.

ID Body black, white tips on hairs; hands and feet whitish. Female short, dark brown ruff with tannish tips; face with diffuse whitish pale brown ring; yellowish malar stripe and moustache. Male paler, black ruff; wrists and ankles brown; vertical line on forehead; white moustache.

HB 39–44 cm
T 45–47 cm
W c. 3 kg

Equatorial Saki *Pithecia aequatorialis* LC
ES **Huapo negro**

EC, PE; W Amazonia.
H Lowland *terra firme* and seasonally flooded white-water forest, S of the lower ríos Napo and Curaray, S to the ríos Bobanana, Conambo and Tigre.

ID Female body grizzled greyish; orange ruff; black face framed with greyish-white; malar stripes, hands, feet white. Male blacker, grizzled white tips to hairs; chest orange; hands and feet white; black face framed in white; white malar stripes, moustache, and white above eyes.

NEOTROPICAL PRIMATES

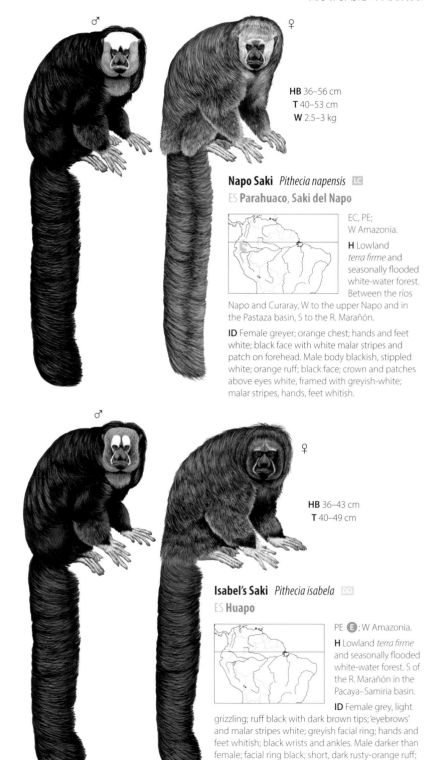

HB 36–56 cm
T 40–53 cm
W 2.5–3 kg

Napo Saki *Pithecia napensis* LC
ES **Parahuaco, Saki del Napo**

EC, PE; W Amazonia.

H Lowland *terra firme* and seasonally flooded white-water forest. Between the ríos Napo and Curaray, W to the upper Napo and in the Pastaza basin, S to the R. Marañón.

ID Female greyer; orange chest; hands and feet white; black face with white malar stripes and patch on forehead. Male body blackish, stippled white; orange ruff; black face; crown and patches above eyes white, framed with greyish-white; malar stripes, hands, feet whitish.

HB 36–43 cm
T 40–49 cm

Isabel's Saki *Pithecia isabela* DD
ES **Huapo**

PE **E**; W Amazonia.

H Lowland *terra firme* and seasonally flooded white-water forest. S of the R. Marañón in the Pacaya–Samiria basin.

ID Female grey, light grizzling; ruff black with dark brown tips; 'eyebrows' and malar stripes white; greyish facial ring; hands and feet whitish; black wrists and ankles. Male darker than female; facial ring black; short, dark rusty-orange ruff; hands, feet, malar stripes, spots above eyes whitish.

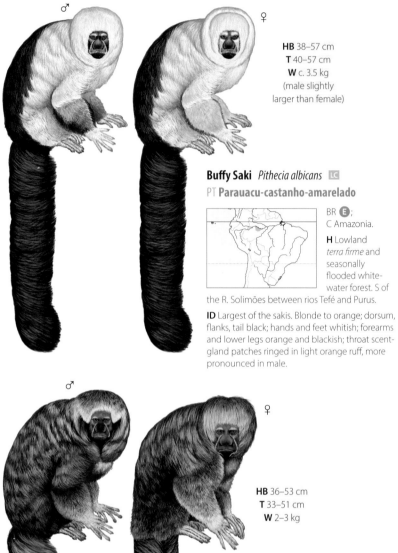

Buffy Saki *Pithecia albicans* LC
PT **Parauacu-castanho-amarelado**

HB 38–57 cm
T 40–57 cm
W c. 3.5 kg
(male slightly larger than female)

BR **E**; C Amazonia.

H Lowland *terra firme* and seasonally flooded white-water forest. S of the R. Solimões between rios Tefé and Purus.

ID Largest of the sakis. Blonde to orange; dorsum, flanks, tail black; hands and feet whitish; forearms and lower legs orange and blackish; throat scent-gland patches ringed in light orange ruff, more pronounced in male.

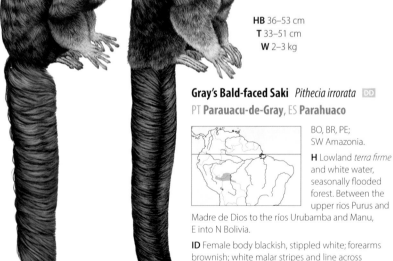

Gray's Bald-faced Saki *Pithecia irrorata* DD
PT **Parauacu-de-Gray**, ES **Parahuaco**

HB 36–53 cm
T 33–51 cm
W 2–3 kg

BO, BR, PE; SW Amazonia.

H Lowland *terra firme* and white water, seasonally flooded forest. Between the upper rios Purus and Madre de Dios to the ríos Urubamba and Manu, E into N Bolivia.

ID Female body blackish, stippled white; forearms brownish; white malar stripes and line across forehead. Male body darker than female; crown, hands, and feet white; face pinkish; dark muzzle outlined by white malar stripes; whitish ruff.

NEOTROPICAL PRIMATES

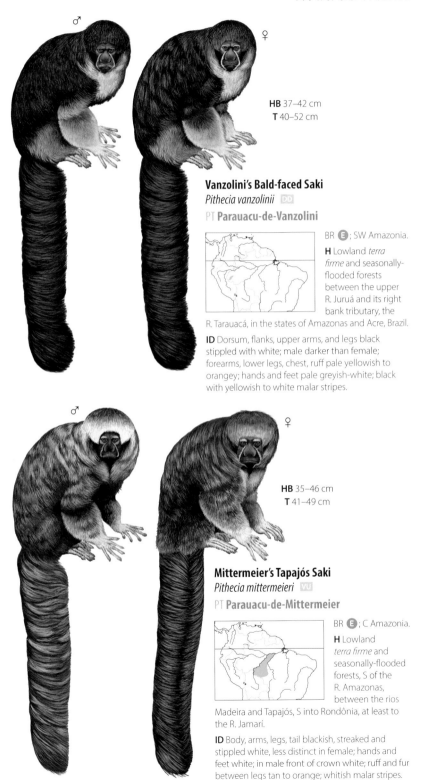

HB 37–42 cm
T 40–52 cm

Vanzolini's Bald-faced Saki
Pithecia vanzolinii DD
PT **Parauacu-de-Vanzolini**

BR **E** ; SW Amazonia.

H Lowland *terra firme* and seasonally-flooded forests between the upper R. Juruá and its right bank tributary, the R. Tarauacá, in the states of Amazonas and Acre, Brazil.

ID Dorsum, flanks, upper arms, and legs black stippled with white; male darker than female; forearms, lower legs, chest, ruff pale yellowish to orangey; hands and feet pale greyish-white; black with yellowish to white malar stripes.

HB 35–46 cm
T 41–49 cm

Mittermeier's Tapajós Saki
Pithecia mittermeieri VU
PT **Parauacu-de-Mittermeier**

BR **E** ; C Amazonia.

H Lowland *terra firme* and seasonally-flooded forests, S of the R. Amazonas, between the rios Madeira and Tapajós, S into Rondônia, at least to the R. Jamarí.

ID Body, arms, legs, tail blackish, streaked and stippled white, less distinct in female; hands and feet white; in male front of crown white; ruff and fur between legs tan to orange; whitish malar stripes.

105

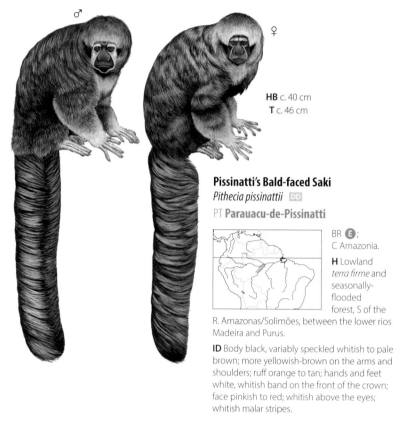

Pissinatti's Bald-faced Saki
Pithecia pissinattii DD
PT **Parauacu-de-Pissinatti**

HB c. 40 cm
T c. 46 cm

BR **E**; C Amazonia.

H Lowland *terra firme* and seasonally-flooded forest, S of the R. Amazonas/Solimões, between the lower rios Madeira and Purus.

ID Body black, variably speckled whitish to pale brown; more yellowish-brown on the arms and shoulders; ruff orange to tan; hands and feet white, whitish band on the front of the crown; face pinkish to red; whitish above the eyes; whitish malar stripes.

Red-nosed Bearded Saki   *Chiropotes albinasus* VU
PT **Cuxiú-de-nariz-vermelho**

BR **E**; E Amazonia.

H *Terra firme* forest, seasonally-flooded forest and forest patches in the forest-cerrado ecotone. Between the rios Madeira and Xingu, S to the R. Guaporé–Mamoré.

ID Black; skin of nose and lips bright red, with short, stiff, whitish or yellowish hairs; both sexes with coronal tufts and beards, more developed in male than female; pink scrotum; bushy black tail.

HB 39–51 cm
T 36–48 cm
W 2.2–3.7 kg

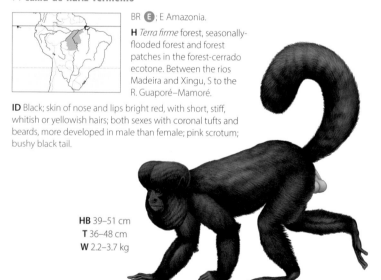

Black Saki *Chiropotes satanas* EN
PT **Cuxiú-preto**

HB 34–42 cm
T 36–42 cm
W 2–4 kg

BR **E**; E Amazonia.

H *Terra firme* forest, secondary forest, forest patches in the cerrado savannah ecotone, and rarely mangrove forest, E from the lower R. Tocantins to the R. Itapecuru basin, Maranhão.

ID Head, beard, tail, underparts black; back and upper limbs dark brown to blackish; facial skin black, sometimes mottled; prominent coronal tufts and beards, more pronounced in male; bright pink scrotum; bushy tail.

△ Up to 200 m

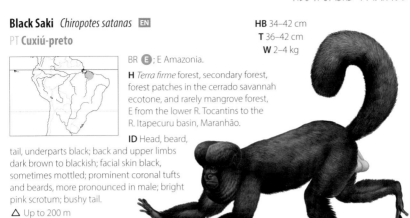

Rio Negro Bearded Saki
Chiropotes chiropotes LC
PT **Cuxiú-do Rio Negro**, ES **Mono barbudo**

HB 35–46 cm
T 30–46 cm
W 2–4 kg

BR, VE; NC Amazonia.

H *Terra firme* and transitional semi-deciduous tropical moist forest, in N Brazil (Amazonas, Roraima) and S Venezuela (Bolívar), delimited by the R. Orinoco in the N, the R. Branco to the E, and the R. Negro in the S.

ID Body brown; dorsum, upper limbs olivaceous; face black, sparsely covered with hair; prominent coronal tufts and beard, more pronounced in male; scrotum and vaginal lips bright pink; bushy tail.

Uta Hick's Bearded Saki
Chiropotes utahicki VU
PT **Cuxiú-de-Uta-Hick**

HB 36–42 cm
T 37–58 cm
W 2–4 kg

BR **E**; E Amazonia.

H Tall, lowland *terra firme* forest and seasonally flooded forest. S of the R. Amazonas, between the rios Xingu and Tocantins–Araguaia, S at least to the R. Taparipe, near the forest-savannah transition.

ID Body predominantly reddish-brown; arms and legs slightly darker than dorsum; prominent coronal tufts and beard, more pronounced in male; bushy tail.

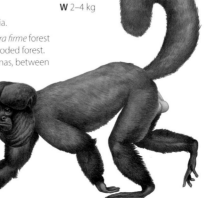

Guianan Bearded Saki
Chiropotes sagulatus LC

PT **Cuxiú, Saki satan**, FR **Saki capucin**, SR **Bisa**

HB 33–46 cm
T 39–46 cm
W c. 3 kg

BR, GF, GY, SR; NE Amazonia.

H *Terra firme* forest, high mountain savannah forest, and seasonally flooded riparian forest of the Guiana Shield, N of the R. Amazonas and E of the Essequibo R. and the rios Branco and Negro.

ID Dorsum and upper limbs orange to reddish-brown; head, nape, lower arms, and legs blackish; prominent coronal tufts and beard, more pronounced in male; bushy tail.

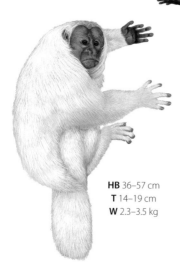

White Bald Uacari *Cacajao calvus* LC
PT **Uacari-branco**

BR E; W Amazonia.

H Seasonally flooded, white-water forest in two areas: the confluence of the rios Solimões and Japurá; and W of the lower R. Juruá, along the right bank of the R. Riozinho, N to the lower R. Jutaí.

ID Dorsum, nape, flanks, tail yellowish to greyish white; underparts, chest, and limbs orange or golden orange; beard dark reddish-brown; face red; prominent swelling on crown.

HB 36–57 cm
T 14–19 cm
W 2.3–3.5 kg

Novaes's Bald Uacari *Cacajao novaesi* VU
PT **Uacari-de Novaes**

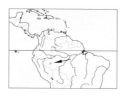

BR E; W Amazonia.

H Seasonally flooded, white-water forest, S of the upper R. Juruá, between the R. Gregório and lower R. Tarauacá.

ID Dorsum, nape, and tail pale orange, buff or whitish; saddle, flanks, underparts, arms, legs reddish-orange; nape and crown whitish; face red; prominent swellings on crown due to hypertrophied temporal muscles.

△ 47–98 m

HB 41.5–49.3 cm
T 14.5–16.5 cm
W c. 3 kg

NEOTROPICAL PRIMATES

HB 36–57 cm
T 14–19 cm
W 2.3–3.5 kg

Red Bald Uacari *Cacajao rubicundus* LC
PT **Uacari-vermelho**

BR **E** ?, CO?; W Amazonia.

H Seasonally flooded, white-water forest, in three areas: along the Jacurapá channel and R. Solimões (in the west); left bank of the lower R. Jutaí, S of the R. Solimões; and along the Auati–Paraná, N of the R. Solimões.

ID Body reddish or reddish-chestnut; nape buffy or whitish; underparts, flanks, limbs reddish or reddish-orange; face red; prominent swellings on crown.

HB 36–57 cm
T 14–19 cm
W 2.3–3.5 kg

Ucayali Bald Uacari *Cacajao ucayalii* VU
PT **Uacari-do-Río-Ucayali**, ES **Huapo rojo**, **Huapo colorado**

BR, PE; W Amazonia.

H *Terra firme* and seasonally flooded forests and palm swamp in the Ucayali–Javari interfluvium and the Pacaya–Samiria basin, with isolated populations to the S and W in the upper Ucayali and montane forest in the upper R. Mayo basin, San Martín, and the Región Junin, in Peru.

ID Body overall reddish-chestnut or reddish-orange; red face; prominent swellings on crown.

△ Up to 2000 m

No measurements available

Kanamari Bald Uacari *Cacajao amuna* NE
PT **Uacari-dos-Kanamari**

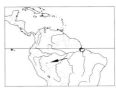

BR **E** ; W Amazonia.

H Seasonally flooded, white-water forest. S of the middle R. Juruá, along E bank of the R. Tarauacá, S to the rios Envira and Jurupari, E to the rios Pauini and Moaco, affluents of the R. Purus.

ID Dorsum and flanks slightly greyish white; nape and upper portion of tail white; arms yellowish-white; inside legs and under tail cream; beard reddish-orange; face red; prominent swellings on crown.

△ 47–98 m

Humboldt's Black-headed Uacari
Cacajao melanocephalus LC

PT **Uacari-preto, Macaco-bicó,** ES **Colimocho, Mono chucuto**

HB 35–56 cm
T 14–16 cm
W 1.9–2.8 kg

BR, CO, VE; NC & W Amazonia.

H Seasonally flooded, black-water forest, *terra firme* and white-sand forest. Between the rios Negro and Japurá/Caquetá, W of the R. Cassiquiare, W to between the Guayabero/Guaviare.

ID Dorsum golden-yellow; lower back redder; reddish-brown thighs, tail darker; nape, mantle, head, beard, arms, hands, and feet blackish; mantle to shoulders with some golden hairs.

Neblina Black-headed Uacari
Cacajao hosomi VU

PT **Uacari-preto,** ES **Mono chucuto**

HB 30–50 cm
T 13–21 cm
W 2.4–4.5 kg
(male larger than female)

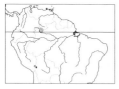

BR, VE; NC Amazonia.

H Lowland permanently or seasonally flooded forest near black-water streams, palm swamp, *terra firme* forest and white-sand, caatinga forest. Between the upper rios Orinoco and Negro.

ID Largest of the black uacaris. Dorsum, tail, thighs bright reddish-brown; nape, mantle, shoulders, head, beard, arms, legs, hands, feet, and tail-tip blackish.
△ 100–1500 m

Ayres's Black-headed Uacari
Cacajao ayresi LC

PT **Uacari-preto- de-Araca**

HB 36.3–39.5 cm
T 15.5–18 cm
W 2.25–24.5 kg

BR E ; NC Amazonia.

H *Terra firme* and permanently or seasonally flooded forest near black-water streams, palm swamp. In the R. Curuduri basin, N of the R. Negro.

ID Small uacari. Dorsum, thighs, tail blackish, with reddish hair showing through; dorsal surface of tail reddish, tip black; dark mantle of longer hairs over shoulders and upper back; underparts, hands, arms, legs (not thighs), and feet black.

NEOTROPICAL PRIMATES

HOWLER MONKEYS, SPIDER MONKEYS, WOOLLY MONKEYS AND MURIQUIS · Atelidae

Colombian Red Howler *Alouatta seniculus seniculus* LC
PT **Guariba-vermelha**, ES **Aullador rojo**, **Coto rojo**, **Mono araguato**

BR, CO, EC, PE, VE; N Andes, W Amazonia.

H Humid, semi-deciduous, deciduous, swamp, seasonally flooded forest and, in the Llanos of E Colombia, gallery forest.

ID Body golden-toned to coppery-red; head, shoulders, limbs, and proximal part of tail, maroon; tail prehensile, gradually paler to tip; male often with blackish beard, limbs and tail; face, black.

△ 0–3200 m

HB 48–63 cm
T 52–80 cm
W 4.1–9 kg
(male larger than female)

Trinidad Red Howler
Alouatta seniculus insulanus NE

TT **E**; Caribbean island.

H Patches of tropical dry evergreen forest, semi-deciduous forest, and swamp/mangrove forests.

ID A rather small howler monkey. Dorsum, head, and flanks red-brown with golden sheen in certain lights; limbs, hands, and feet bright red-brown deepening to maroon on forearms; tail prehensile, base maroon shading to golden distally.

HB 43–52 cm
T 53–61 cm
W 4–6 kg
(male slightly larger than female)

Juruá Red Howler
Alouatta juara LC

PT **Guariba-vermelha-do-Juruá**, ES **Araguato**, **Coto**

BR, PE; W Amazonia.

H Tall tropical rain forest, swamp and seasonally flooded forest. Sometimes rare or absent in Amazonian *terra firme* forest. South of the R. Solimões, W of the R. Purus into E Peru.

ID Not sexually dichromatic. Body dark reddish-brown with small, orangey-rufous area in mid-dorsum; beard, cheeks, limbs, and base of tail dark blackish-rufous, paler on hands and feet; tail more golden distally.

HB c. 52 cm
T c. 62 cm

111

Purus Red Howler *Alouatta puruensis* VU
PT **Guariba-vermelha-do-Purus**, ES **Araguato, Coto**

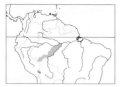

BR, PE; W Amazonia.

H Tall forest, swamp and seasonally flooded forest. In the R. Purus basin, E to the lower and middle R. Madeira.

ID Sexually dichromatic. Male dark reddish-orange, paler on dorsum; female pale golden-orange on dorsum and flanks; crown, nape, forearms, tail base, and beard darker, reddish to blackish mahogany; tail paler towards tip.

HB 50.5–67 cm
T 59–65.5 cm

Ursine Red Howler *Alouatta arctoidea* LC
ES **Mono araguato**

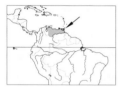

VE **E**; N Andes, Llanos, Caribbean coast.

H Deciduous forest patches, open woodland, gallery forest, cloud forest. E from the Cordillera de Merida, N of the ríos Orinoco and Apure.

ID Body golden-toned to coppery-red; head, shoulders, limbs, and proximal part of tail maroon; sexes similar but male often has blackish beard, limbs, and tail.

△ 10–2000 m

HB 45–65 cm
T 55–68 cm
W 4.5–8 kg
(male larger than female)

Guianan Red Howler *Alouatta macconnelli* LC
PT **Guariba-vermelha**, ES **Mono araguato**,
FR **Hurleur roux**, SR **Baboen**

BR, GF, GY, SR, VE; NE Amazonia.

H High rainforest, marsh and swamp forest, gallery forest and forest patches in savannahs. rarely in liana forest, swamp scrub, and mangrove, south of the R. Orinoco.

ID Body uniformly dark rufous-brown, with yellowish-brown to golden back; dark dorsal stripe; head rich maroon; forearms and lower legs orange-red; prehensile tail gradually paler distally.

△ Up to 1200 m

HB 48–63 cm
T 52–80 cm
W 4–9 kg
(male larger than female)

Bolivian Red Howler *Alouatta sara* NT
PT **Guariba-vermelha**, ES **Manechi Colorado, Coto, Mono aullador**

BO, BR, PE; S & W Amazonia.

H *Terra firme* and seasonally flooded forest. In the R. Beni basin, Bolivia, extending NE, E of the R. Madeira through S Rondônia state, to the rios Aripuanã and probably Juruena, Mato Grosso, W along the Madre de Dios basin to the R. Inuya, Peru.

ID Body brick-red, with limbs, head, and proximal part of tail slightly darker (more rufous); blackish fringe around face.

HB 54–71.2 cm
T 51.5–60 cm
W 6–9 kg

Amazon Black Howler *Alouatta nigerrima* LC
PT **Guariba-preta-da-Amazônia**

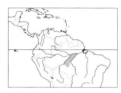

BR E ; E Amazonia.

H *Terra firme* and seasonally inundated forest, forest patches in savannahs and secondary forest. Between the lower and middle rios Madeira and Tapajós, with outlying populations around Oriximiná and Obidos, N of the R. Amazonas and the Lago Janauacá, W of the lower R. Madeira.

ID A large, entirely black species, with minimal sexual dichromatism.

HB 48.5–64.8 cm
T 56.5–69 cm
W 4.9–8 kg
(male larger than female)

Red-handed Howler *Alouatta belzebul* VU
PT **Guariba-preta-de-mãos-ruivas**

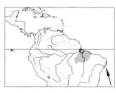

BR E ; E Amazonia, N Atlantic Forest.

H *Terra firme* forest, flooded, seasonally flooded and semi-deciduous forest. S of the R. Amazonas, E from the rios Xingu and Iriri, islands in the delta. Also, few locations in cerrado, caatinga and babassu palm forest in NE Brazil.

ID Largely black. Hands, feet, tail tip, and sometimes forehead and back reddish-brown to yellow; scrotum red.

HB 37–65 cm
T 45–70 cm
W 4.8–8 kg
(male larger than female)

Spix's Howler *Alouatta discolor* VU
PT Guariba-preta

BR **E**; E Amazonia.

H *Terra firme* and seasonally flooded forest, palm swamp. S of the R. Amazonas, E from the rios Tapajós and Juruena to the Xingu and Iriri, S to the R. Santa Helena, tributary of the R. Teles Pires, Mato Grosso, and the Serra do Cachimbo, Pará.

ID Dark brown to black; dorsal band, rump, flanks, thighs, inner arms, hands, feet, and tip of tail rufous-chestnut.

HB 46.5–91.5 cm
T 60.5–67.5 cm
W 4.8–8 kg
(male larger than female)

Maranhão Red-handed Howler *Alouatta ululata* EN
PT Guariba-da-Caatinga

BR **E**; E Amazonia.

H Forest patches along the coast, semi-deciduous forests in caatinga (dry forest scrub), transitional Babassu palm (*Orbygnia*) forests in Maranhão, and mangroves on the coasts of Piauí and Maranhão.

ID Sexually dichromatic. Male black, with rufous to reddish-brown hands and feet, tail tip, and flanks. Female yellowish-brown with sparse greyish hairs, giving an olivaceous appearance.

HB c. 56.5 cm
T c. 56 cm
(from adult male type)

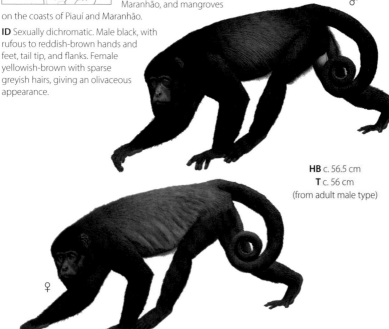

Brown Howler *Alouatta guariba* VU
PT **Bugio**, **Bugio-ruivo**, **Barbado**

AR, BR; Atlantic Forest.

H Lowland, submontane, evergreen, and semi-deciduous forests.
S from the R. de Contas, Bahia, to *Araucaria* pine forest in the S, and just into Misiones, Argentina.

ID Male variably orange-red to red-brown with a golden sheen; large beard. Female varies from brown to dark brown or blackish with dorsum frosted orange or yellow-brown hairs.

△ 0–700 m

variants

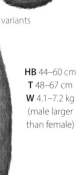

HB 44–60 cm
T 48–67 cm
W 4.1–7.2 kg
(male larger than female)

Black and Gold Howler *Alouatta caraya* NT
PT **Bugio-preto**, ES **Carayá negro y dorado**, **Manechi negro**

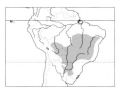

AR, BO, BR, PY, UY; Caatinga, Cerrado, Chaco, S Amazonia.

H Forest patches in savannah, gallery forest, deciduous and semi-deciduous forest, Chaco, and riparian and seasonally flooded forest of the Pantanal.

ID Young of both sexes blond. Mature male black, but in some regions partly brown to yellowish-brown; female pale yellow (straw) or golden-brown.

HB 50–65 cm
T 54.5–65 cm
W 3.6–9.6 kg
(male larger than female)

Golden-mantled Howler *Alouatta palliata palliata* EN
ES **Mono congo, Mono aullador, Mono Zaraguate**

CR, GT, HN, NI, PA; C America.

H Lowland evergreen, *terra firme*, riparian and seasonally-flooded forest, sub-montane and semi-deciduous forest. In C Guatemala (R. Motagua to Cabo de Tres Puntas to meet *A. pigra*), E Costa Rica, and W Panama.

ID Coat silky black; mantle and flanks extending to dorsum with longer, gold/yellowish-brown fur, variable in colour and extent; white scrotum.

△ 0–2500 m

HB 46–63 cm
T 55–70 cm
W 3.1–9 kg
(male larger than female)

Azuero Peninsula Howler *Alouatta palliata trabeata* EN
ES **Mono gun gun**

PA **E**; C America.

H Remnant forest patches on hill tops, and gallery forest in the provinces of Herrera, Los Santos and Veraguas, on the Azuero Peninsula, SW Panama.

ID Pelage brownish-black; head, shoulders, upper back, tail, and distal part of limbs walnut brown; golden tinge to flanks and loins.

△ Up to 1500 m

HB 46–63 cm
T 55–70 cm
W 3.1–9 kg
(male larger than female)

Coiba Island Howler *Alouatta palliata coibensis* EN
ES **Mono aullador de Coiba**

PA **E**; C America.

H Humid *terra firme* and swamp forest on the islands of Coiba and Jicarón in the Gulf of Chiriqui, off the west coast of Panama.

ID Head and upper back dark seal brown; lower back paler; flank hairs elongated, orange-rufous to cinnamon rufous; veil more restricted to flanks than in *A. p. trabeata*; rump and hindlimb walnut-brown.

HB c. 56 cm (M)
T c. 58 cm (M)
(male larger than female)

NEOTROPICAL PRIMATES

Ecuadorian Mantled Howler *Alouatta palliata aequatorialis* VU
ES **Aullador negro, Mono coto de Tumbes**

CO, EC, PA, PE; Pacific coast, E of Andes, C America.

H Foothills, lowlands, and lower montane forests. From W Panama in the Darién to W Colombia, N through the basins of the ríos Sinú and Atrato, and S, W of the Andes to NW Peru.

ID Black to dark brown; mantle hairs shorter than those of *A. p. palliata*; sparse golden-ochraceous hairs on back; pendulous white scrotum.

HB 48.5–61.5 cm
T 56–65 cm
W 3.1–9 kg
(male larger than female)

Mexican Mantled Howler *Alouatta palliata mexicana* EN
ES **Zaraguate, Saraguato café**

MX.

H Remnant lowland and sub-montane forest patches in S and SE Mexico.

ID Pelage blackish-brown; head, shoulders, limbs, and tail black; light banded, silvery hairs more widely distributed over back than in *A. p. palliata*; pendulous white scrotum.

HB 33–45.5 cm
T 50–69 cm
W 3.60–7.20 kg
(male larger than female)

Central American Black Howler *Alouatta pigra* EN
ES **Saraguato negro, Mono Zaraguate; Baboon (BZ)**

BZ, GT, MX; C America.

H Lowland *terra firme*, riparian and seasonally-flooded forest (Yucatán moist forest), sub-montane and semi-deciduous forest. In SE Mexico, Belize, and N & C Guatemala, S to the Lago de Izabal, El Golfete, and R. Dulce in Guatemala.

ID Body with long, soft, dense, black fur; traces of brown on shoulders, cheeks, and back; small crest on crown; white scrotum.

Δ 0–3350 m

HB 34.5–54 cm
T 56–70.7 cm
W 44.5–9.6 kg
(male larger than female)

117

Geoffroy's Spider Monkey
Ateles geoffroyi geoffroyi CR
ES **Mono colorado**

NI E ; C America.

H Lowland and submontane forest patches. From the coast of SE Nicaragua, NW, W of the Cordillera Chantaleña to Lake Managua.

ID Back, upper arms, thighs yellowish-buff, silvery to brownish-grey; underparts, elbows, forearms, knees, lower legs, hands, feet black; crown dark, with mixed light and dark hairs directed forward; face black; pink 'spectacles'; tail like back.

HB 31–63 cm
T 64–86 cm
W 6–9 kg

Azuero Spider Monkey
Ateles geoffroyi azuerensis CR
ES **Mono charro**

PA E ; C America.

H Forested mountains of the W side of the Azuero Peninsula (Veraguas Province) in the vicinity of Ponuga, Panama.

ID Body light tawny or ochraceous tawny; surfaces of limbs black; crown black or blackish-brown; lateral surfaces of arms and legs black; throat and chest cinnamon buff; face blackish; tail black above, tawny below.

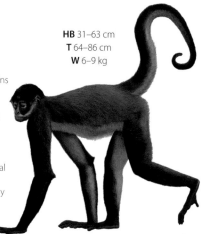

HB 31–63 cm
T 64–86 cm
W 6–9 kg

Black-browed Spider Monkey
Ateles geoffroyi frontatus VU
ES **Mono colorado**

CR, NI; C America.

H Lowland rainforest, semi-deciduous and deciduous forest. In N and W Nicaragua, along the Pacific coast, W of the Guanacaste Mts. to the Nicoya Peninsula, NW Costa Rica.

ID Similar to *A. g. geoffroyi* in restriction of black to crown and lateral aspects of limbs but darker overall; upperparts brownish; underparts honey yellow to tawny; tail blackish above, tawny below.
Δ 0–1800 m

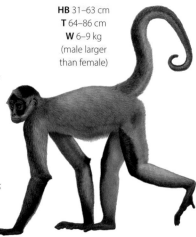

HB 31–63 cm
T 64–86 cm
W 6–9 kg
(male larger than female)

Hooded Spider Monkey
Ateles geoffroyi grisescens DD

ES **Mono araña gris del Darién**

No measurements available

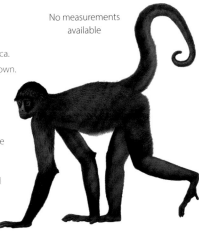

CO(?), PA(?); C America.

H Provenance unknown. Possibly montane and pre-montane forests of the Darién Gap in the valley of the R. Tuyra, SE via the Serranía del Sapo to extreme SE Panama and the Cordillera de Baudó, NW Colombia. See page 20.

ID Pelage moderately long, black, with many silvery-white hairs interspersed; hair of the forehead moderately long; tail black, underside greyish.

Ornate Spider Monkey
Ateles geoffroyi ornatus VU

ES **Mono colorado**

HB 31–63 cm
T 64–86 cm
W 6–9 kg

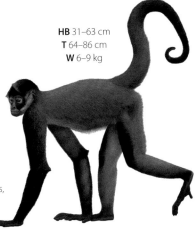

CR, NI; C America.

H Primary and secondary lowland rainforest; evergreen, semi-deciduous and cloud forest. In C and E Costa Rica, and Panama from Chiriquí Province to the Serranía de San Blas, E of the Canal Zone.

ID Upperparts red-brown to dark golden brown; head, face, crown, forearms, outer sides of legs, hands, feet black; underside does not contrast strongly with dorsum; tail black above, tawny below.

Mexican Spider Monkey
Ateles geoffroyi vellerosus EN

ES **Mono colorado; Monkey (BZ)**

HB 31–63 cm
T 64–86 cm
W 6–9 kg

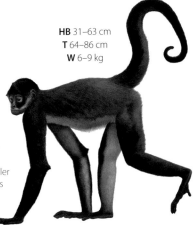

BZ, SV, GY, HN, MX; C America.

H Primary and secondary lowland rainforest; dry, semi-deciduous and cloud forest, and mangrove swamps. In E and SE Mexico, Yucatán Peninsula, Belize, Guatemala, El Salvador and Honduras.

ID Upperparts from black to dark brown, except for paler band across lumbar region; underparts and inner limbs paler; pink around eyes; crown, neck, shoulders, arms, legs, and tail brownish-black.

△ 0–2000 m

Colombian Black Spider Monkey   *Ateles rufiventris*   VU
ES **Marimonda chocoana**, **Mono araña negro del Darién**

CO, PA; W of N Andes.

H Dry, humid, and cloud forest along the Cordillera Occidental, Colombia. W from the ríos Cauca and Magdalena, S from the Canal del Dique to Barbacoas, R. Telembi, Nariño, NW into SE Panama in the provinces of Darien and Panama.

ID Glossy-black; slight brownish tinge on forehead; cheeks and muzzle with sparse whitish or golden hairs; face black; genital area to inner thigh hairs reddish.

△ 50–2500 m

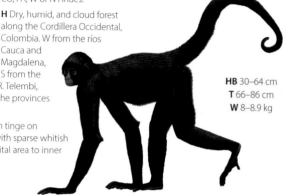

HB 30–64 cm
T 66–86 cm
W 8–8.9 kg

Ecuadorian Spider Monkey   *Ateles fusciceps*   CR
ES **Mono araña de cabeza marrón**

EC **E**; W of N Andes, Pacific Coast.

H Patches of tropical lowland and subtropical humid forests. In the provinces of Manabi, Esmeraldas, Carchi, Pichincha and possibly Santo Domingo de los Tsachilas in NW Ecuador.

ID Coat black to brownish-black; anterior crown rust brown to yellowish-brown, with a low, point tuft; crown grades from brown to black on nape; chin and lips with short white hairs; face black.

△ 300–2500 m

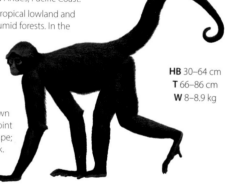

HB 30–64 cm
T 66–86 cm
W 8–8.9 kg

Black Spider Monkey   *Ateles chamek*   EN
PT **Coatá-de-cara-preta**, ES **Maquisapa**, **Mono araña**

HB 40–60 cm
T 70–88 cm
W 7–9 kg

BO, BR, PE; S & W Amazonia.

H Primary lowland rainforest, semi-deciduous, riparian and flooded forest, S of the R. Amazonas/Solimões, W of the rios Tapajós–Teles Pires to the R. Ucayali and, S of the R. Cushabatay, the R. Huallaga, S to N and C Bolivia.

ID Black like *A. paniscus* but smaller and shorter-haired; largely black, not red, facial skin; muzzle, cheeks, and forehead sometimes with a few white hairs.

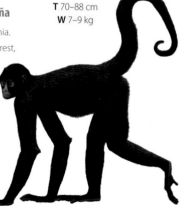

Red-faced Black Spider Monkey
Ateles paniscus VU

PT **Coatá-de-cara-vermelha**, FR **Atèle noir**,
SR **Kwatta**

HB 51.5–66 cm
T 64–93 cm
W 9.6–10.8 kg

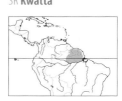

BR, GF, GY, SR; NE Amazonia.

H High forest, rarely in edge or degraded forest and *Euterpe* palm swamp on the Guiana Shield, E of the Essequibo R., Guyana, excluding lowland coastal plains, N of the R. Amazonas, E of the rios Negro and Branco.

ID Long, silky, glossy-black fur; face naked, pink or reddish; tail thickly furred for two-thirds of length, then tapering sharply toward tip.

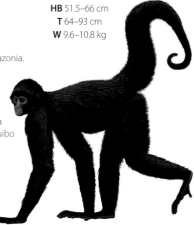

White-whiskered Spider Monkey
Ateles marginatus EN

PT **Coatá-de-cara-branca**,
Macaco-aranha-de-testa-branca

HB 37–70 cm
T 62–90 cm
W 5.8–9 kg

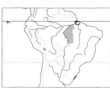

BR E ; SE Amazonia.

H Primary lowland rain forest, and riparian, marsh, semi-deciduous, and dry savannah forest, S of the lower R. Amazonas between the rios Tapajós and Teles Pires and the R. Xingu (left bank), S to the Serra do Cachimbo.

ID Fur entirely black except triangular white patch on forehead and whitish cheek whiskers. Juveniles with pale, flesh-coloured face that becomes red in adults.

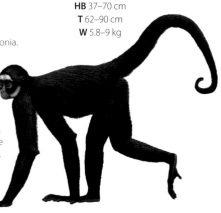

White-bellied Spider Monkey
Ateles belzebuth EN

PT **Coatá-de-barriga-branca**,
ES **Mono araña**, **Marimonda**, **Maquisapa**

variants

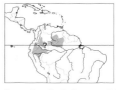

BR, CO, EC, PE, VE; NW Amazonia.

H Montane, submontane, lowland riparian, marsh, and semi-deciduous forest. E to the R. Caura, in Colombia N to the R. Upía, Brazil to the R. Japurá, N of the R. Negro to the R. Branco, S of the R. Napo in Ecuador and Peru, S to the R. Cushabatay.

HB 46–50 cm
T 74–81 cm
W ≤ 10 kg

ID Body blackish; yellowish-white underparts, inner limbs, and underside of tail; many with triangular white or golden patch on forehead.

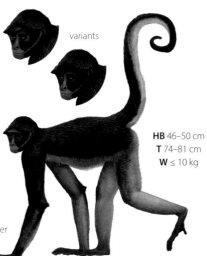

Variegated Spider Monkey *Ateles hybridus* CR
ES **Mono araña de vientre amarillo, Marimonda**

CO, VE; N & E Andes.
H Evergreen, semi-deciduous, riparian, and montane forests. In N Colombia, lowlands between the lower ríos Cauca and Magdalena, NE to the R. Perijá into Catatumbo and Arauca, E into Venezuela.

ID Upperparts and hind limbs pale brown; underparts, inner limbs, tail whitish to buffy or pale yellow; head and forearms darker brown; white patch on forehead; ventral surface of tail paler.

△ 20–1500 m

HB 44–50 cm
T 74–81 cm
W 7.5–10.5 kg

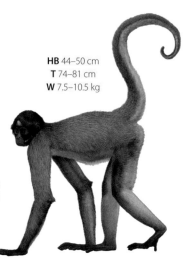

Humboldt's Woolly Monkey
Lagothrix lagothricha lagothricha VU
PT **Macaco-barrigudo**, ES **Churuco, Mono lanudo**

BR, CO, EC, PE; NW Amazonia.
H Lowland *terra firme* and seasonally-flooded forest From the Cordillera Oriental, N to the R. Guaviare, E to the Orinoco, N of the ríos Aguarico, Napo and Amazonas/Solimões, E to the mouth of the R. Içá.

ID Coats variable, from very light blond to greyish-brown, dark brown or blackish; head same shade as back or paler; limbs and tail darker than back.

△ 30–350 m

HB 46–65 cm
T 53–77 cm
W 5–7 kg
robust dominant males up to 10 kg

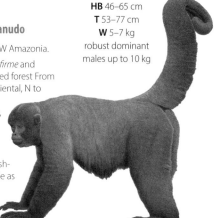

Colombian Woolly Monkey
Lagothrix lagothricha lugens CR
ES **Churuco, Mono lanudo**

CO (E), E of N Andes.
H Lowland, submontane and montane forests of the Cordillera Oriental in the upper Magdalena basin and the Serranía de San Lucas, N from the R. Caquetá to the upper R. Arauca, Colombia.

ID Head, arms, and body pale grey to dark brown or blackish, the last typical at higher altitudes; legs and tail blackish-brown with some grey hairs.

△ 200–2500 m

HB 46–65 cm
T 53–77 cm
W 5–7 kg
robust dominant males up to 10 kg

Grey Woolly Monkey
Lagothrix lagothricha cana EN
PT **Macaco-barrigudo**

BR **E** ; C Amazonia.

H *Terra firme* forest, seasonally flooded and semi-deciduous forests of the R. Solimões between the R. Juruá and rios Tapajós and Juruena, S to the rios Madeira and Abunã and below the mouth of the R. Jiparaná, extending E to the Tapajós.

ID Coat brownish-grey, olivaceous, often with dark mid-dorsal streak; head darker; tail coloured like back.

△ 25–300 m

HB 42–58 cm
T 61–69 cm
W c. 7 kg
robust dominant males up to 10 kg

Poeppig's Woolly Monkey
Lagothrix lagothricha poeppigii EN
PT **Macaco-barrigudo**, ES **Mono lanudo rojizo**, **Choro**, **Chorongo**

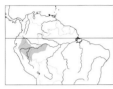

BR, EC, PE; W Amazonia.

H Lowland *terra firme* riparian and seasonally-flooded and sub-montane forest from the Cordillera Oriental, S of the rios Amazonas/Solimões and Napo in Ecuador, Peru and Brazil, E to the R. Juruá, S to the R. Pachitea in Peru.

ID Coat ochraceous-buff to chestnut; flanks reddish; crown dark brown; forearms and forelegs brown; underparts deep reddish; head, hands, feet black.

△ 35–1000 m

HB 46–56 cm
T 55–67.5 cm
W 5–7 kg
robust dominant males up to 10 kg

Peruvian Woolly Monkey
Lagothrix lagothricha tschudii DD
ES **Choro**, **Mono lanudo**

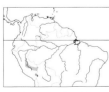

BO, PE; SW Amazonia.

H Montane and submontane Andean forests in SE Peru between the ríos Pachitea and Ucayali, S to the R. Inambari, E through the Madre de Dios and Tambopata basins to the Cordillera de Apolobamba, and Madidi, Bolivia.

ID Coat dark grey, darker and with longer hair than in *L. l. cana*, with tinge of red; dark dorsal streak; head, limbs, and tail black.

△ 150–3000 m

HB 40–65 cm
T 53–80 cm
W 5–7 kg
robust dominant males up to 10 kg

Peruvian Yellow-tailed Woolly Monkey
Lagothrix flavicauda CR
ES **Mono choro de cola amarilla**

HB 40–56 cm
T 56–63 cm
W 5.7–10 kg

PE **E**; C Andes.
H Montane dry forest and cloud forest between the ríos Marañón and Huallaga in the regions of Amazonas, San Martín, E Loreto, Huánuco and La Libertad, S to the Pampa Hermosa basin, Junin.
ID Coat long, thick, deep mahogany; darker lower back, nape, and extremities; distal half of tail yellow; face brown; buffy triangular patch on muzzle; yellow scrotal tuft, female with smaller tuft around vulva.
△ 1400–2800 m

Southern Muriqui
Brachyteles arachnoides CR
PT **Muriqui-do-sul, Mono carvoeiro**

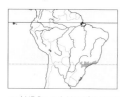

BR **E**; Atlantic Forest.
H Montane, submontane, semi-deciduous and deciduous forest in the Serra do Mar in the states of Rio de Janeiro, São Paulo and NE Paraná, N to the Serra da Mantiqueira, S to the rios Paraíba and Paraíba do Sul.
ID Coat variable, beige, pale brown or light grey-brown; face, palms, and soles black, with minor depigmentation in adults; pronounced belly; no external thumb; prehensile tail.
△ 40–1500 m

HB c. 48 cm
T c. 74 cm
W 8–12 kg

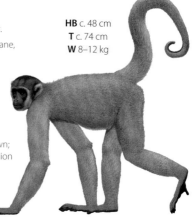

Northern Muriqui
Brachyteles hypoxanthus CR
PT **Muriqui-do-norte**

BR **E**; Atlantic Forest.
H Evergreen, semi-deciduous and deciduous forest patches in the states of Bahia, Espírito Santo, Minas Gerais, and Rio de Janeiro, S to the Serra da Mantiqueira.
ID Coat predominantly beige, pale to dark brown or pale grey-brown; face and genitalia black when young but considerably depigmented when adult; whitish cheeks, forehead, and muzzle; vestigial thumb; pronounced belly; prehensile tail.
△ 250–1350 m

HB c. 60 cm
T 73–81 cm
W c. 12 kg

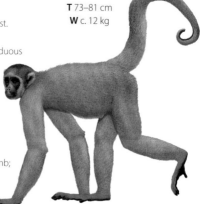

INTRODUCED OLD WORLD MONKEYS · Cercopithecidae

Green Monkey *Chlorocebus sabaeus* LC

BB, KN, LC. Originally from W Africa.

H Montane forest, forest edge, moist deciduous and semi-deciduous forest.

ID Grizzled golden-green above; creamy white undersides and tail; back of thighs yellow; hands and feet grey; face black; yellow whiskers; light browband; tail-tip golden-yellow.

HB 42–60 cm (M)
30–50 cm (F)
T 46–76 cm (M)
41–66 cm (F)
W 3.1–6.4 kg (M)
1.5–4.9 kg (F)

Vervet Monkey *Chlorocebus pygerythrus* LC

AI, MF, SX. Originally from E Africa.

H Montane forest, forest edge, moist deciduous and semi-deciduous forest.

ID Grizzled-grey or olive above, including crown and outer surfaces of limbs; undersides, browband, and short whiskers white; face black; dark hands, feet, and tail tip; scrotum turquoise.

HB 42–70 cm (M), 30–62 cm (F)
T 45–76 cm (M), 41–66 cm (F)
W 3.1–6.4 kg (M), 1.5–4.9 kg (F)

Mona Monkey *Cercopithecus mona* NT

GD.

H Montane, moist deciduous and semi-deciduous forest. Originally from W Africa.

ID Chestnut back; white undersides; crown grey speckled with yellow-gold; puffy cheeks, browband white; muzzle pink; dark upper nose and around eyes; dark hair on temples; tail base with white patches; outside lower limbs dark; callosities dark grey; scrotum blue.

△ 220–710 m

HB 41–63 cm (M)
38–46 cm (F)
T 64–88 cm (M)
53–66 cm (F)
W 4.4–7.5 kg (M)
2.5–4 kg (F)

Checklist of the Neotropical Primates

IUCN Red List conservation status
NE Not Evaluated · DD Data Deficient · LC Least Concern · NT Near Threatened · VU Vulnerable · EN Endangered · CR Critically Endangered

MARMOSETS, GOELDI'S MONKEY, TAMARINS AND LION TAMARINS · CALLITRICHIDAE · p. 46

Northern Pygmy Marmoset — *Cebuella pygmaea* — VU — Brazil, Colombia, Ecuador, Peru
Mono de Bolsillo (BO), Mico-leãozinho (BR), Leoncillo (CO), Leoncito (CO, PE), Leoncillo del norte (EC), Tití pigmeo del norte (EC)

Southern Pygmy Marmoset — *Cebuella niveiventris* — VU — Bolivia, Brazil, Ecuador, Peru
Tití pigmeo (BO), Mono de Bolsillo (BO, PE), Mico-leãozinho (BR), Leoncito (CO, PE), Leoncillo del sur (EC, PE), Tití pigmeo del sur (EC), Tití enano (PE)

Black-crowned Dwarf Marmoset — *Callibella humilis* — LC — Brazil
Sauim-anão (BR), Soim (BR), Suim (BR)

Silvery Marmoset — *Mico argentatus* — LC — Brazil
Souim-argênteo (BR), Choim (BR), Suim (BR)

Golden-white Bare-ear Marmoset — *Mico leucippe* — LC — Brazil
Souim-branco (BR), Choim (BR), Suim (BR)

Snethlage's Marmoset — *Mico emiliae* — LC — Brazil
Mico-de-Emilia (BR), Souim (BR), Choim (BR), Suim (BR)

Munduruku Marmoset — *Mico munduruku* — VU — Brazil
Souim-dos-Munduruku (BR), Choim (BR), Suim (BR)

Schneider's Marmoset — *Mico schneideri* — EN — Brazil
Souim-de-Schneider (BR), Choim (BR), Suim (BR)

Black-and-white Tassel-ear Marmoset — *Mico humeralifer* — NT — Brazil
Santarém Marmoset; Sagüi-de-Santarém (BR), Souim (BR), Choim (BR), Suim (BR)

Golden-white Tassel-ear Marmoset — *Mico chrysoleucos* — LC — Brazil
Souim-dourado-e-branco (BR), Choim (BR), Suim (BR)

Maués Marmoset — *Mico mauesi* — LC — Brazil
Souim-de-Maués (BR), Choim (BR), Suim (BR)

Sateré Marmoset — *Mico saterei* — LC — Brazil
Souim-de-Sateré (BR), Choim (BR), Suim (BR)

Rio Acarí Marmoset — *Mico acariensis* — LC — Brazil
Souim-do-rio-Acari (BR), Choim (BR), Suim (BR)

Black-headed Marmoset — *Mico nigriceps* — NT — Brazil
Souim-de-cabeça-preta (BR), Soim (BR), Suim (BR)

Marca's Marmoset — *Mico marcai* — VU — Brazil
Souim-de-Marca (BR), Choim (BR), Suim (BR)

Rio Aripuanã Marmoset — *Mico intermedius* — LC — Brazil
Souim-do-rio-Aripuanã (BR), Choim (BR), Suim (BR)

Black-tailed Marmoset — *Mico melanurus* — NT — Bolivia, Brazil, Paraguay
Leoncito (BO), Tití de cola negra (BO, PY), Mico-de-rabo-preto (BR), Souim-de-rabo-preto (BR), Choim (BR), Suim (BR), Leoncito (BO), Mono eléctrico (PY)

NEOTROPICAL PRIMATES

Rondon's Marmoset — *Mico rondoni* — VU — Brazil
Sauim-de-Rondônia (BR), Soim (BR), Suim (BR)

Goeldi's Monkey — *Callimico goeldii* — VU — Bolivia, Brazil, Colombia, Peru
Mico de Goeldi (BO), Mico-de-Goeldi (BR), Taboqueiro (BR), Sauim-preto (BR), Diablillo (CO), Chichico negro (CO), Supay pichico (PE), Maquisapita (PE), Leoncillo (PE), Pichico (PE), Pichico falso de Goeldi (PE)

White-tufted-ear Marmoset — *Callithrix jacchus* — LC — Brazil
Sagüi-de-tufo-branco (BR), Sagüi-do-nordeste (BR)

Black-tufted-ear Marmoset — *Callithrix penicillata* — LC — Brazil
Mico-estrela (BR), Sagüi-de-tufo-preto (BR)

Wied's Black-tufted-ear Marmoset — *Callithrix kuhlii* — VU — Brazil
Sagüi-de-Wied (BR)

White-faced Marmoset — *Callithrix geoffroyi* — LC — Brazil
Sagüi-da-cara-branca (BR)

Buffy-headed Marmoset — *Callithrix flaviceps* — CR — Brazil
Sagüi-da serra (BR), Sagüi-da-serra-claro (BR), Sagüi-taquara (BR)

Buffy-tufted-ear Marmoset — *Callithrix aurita* — EN — Brazil
Sagüi-da-serra-escuro (BR), Sagüi-caveirinha (BR)

Spix's Black-mantled Tamarin — *Leontocebus nigricollis nigricollis* — LC — Brazil, Colombia, Peru
Sauim-de-manto-preto (BR), Bebeleche (CO), Boquiblanco (CO), Pichico barba blanca (PE)

Graells's Black-mantled Tamarin — *Leontocebus nigricollis graellsi* — NT — Colombia, Ecuador, Peru
Bebeleche (CO), Boquiblanco (CO), Chichico negro (EC), Tamarin de dorso negro (EC), Chichico del Napo (EC), Pichico barba blanca (PE)

Hernández-Camacho's Black-mantled Tamarin — *Leontocebus nigricollis hernandezi* — LC — Colombia
Bebeleche (CO), Boquiblanco (CO)

Lesson's Saddle-back Tamarin — *Leontocebus fuscus* — LC — Brazil, Colombia
Sauim-de-costas-malhadas-de-Lesson (BR), Soim (BR), Suim (BR), Bebeleche (CO), Pichico (CO)

Golden-mantled Saddle-back Tamarin — *Leontocebus tripartitus* — NT — Ecuador, Peru
Chichico dorado (EC), Chichico de manto dorado (EC), Tamarin de dorso dorado (EC), Pichico dorado (PE)

Red-mantled Saddle-back Tamarin — *Leontocebus lagonotus* — LC — Ecuador, Peru
Chichico rojo (EC), tamarín de dorso rojo (EC), Pichico común (PE)

Andean Saddle-back Tamarin — *Leontocebus leucogenys* — LC — Peru
Pichico común (PE), Pichico andino (PE)

Illiger's Saddle-back Tamarin — *Leontocebus illigeri* — NT — Peru
Pichico común (PE)

Geoffroy's Saddle-back Tamarin — *Leontocebus nigrifrons* — LC — Peru
Pichico común (PE)

Spix's Saddle-back Tamarin — *Leontocebus fuscicollis* — LC — Brazil, Peru
Tamarino de cabeza amarilla (BO), Leoncito (BO). Sauim-de-costas-malhadas-de-Spix (BR), Soim (BR), Suim (BR), Pichico (PE), Pichico común (PE), Pichico de barba blanca (PE)

Cruz Lima's Saddle-back Tamarin — *Leontocebus cruzlimai* — LC — Brazil
Sauim-de-Cruz-Lima (BR), Soim (BR), Suim (BR)

Ávila-Pires's Saddle-back Tamarin	*Leontocebus avilapiresi*	LC	Brazil

Sauim-de-costas-malhadas-de-Ávila-Pires (BR), Soim (BR), Suim (BR)

Grey-fronted Saddle-back Tamarin	*Leontocebus mura*	NT	Brazil

Sauim (BR), Soim (BR), Suim (BR)

Hershkovitz's Saddle-back Tamarin	*Leontocebus primitivus*	DD	Brazil

Sauim-de-costas-malhadas-de-Hershkovitz (BR), Soim (BR), Suim (BR)

Weddell's Saddle-back Tamarin	*Leontocebus weddelli*	LC	Bolivia, Brazil, Peru

Leoncito, Chichilo común (BO), Sauim-costas-malhadas-de-Weddell (BR), Soim (BR), Suim (BR)

White Saddle-back Tamarin	*Leontocebus melanoleucus*	LC	Brazil, Peru

Sauim-branco (BR), Soim-branco (BR), Suim (BR), Pichico blanco (PE)

Spix's Moustached Tamarin	*Tamarinus mystax*	LC	Brazil, Peru

Sauim-de-bigode (BR), Bigodeiro (BR), Pichico barba blanca (PE), Cervecero (PE)

Red-capped Moustached Tamarin	*Tamarinus pileatus pileatus*	LC	Brazil

Sauim (BR), Sauim-de-boca-branca (BR), Soim (BR)

White-rumped Moustached Tamarin	*Tamarinus pileatus pluto*	LC	Brazil

Sauim-de-bigode (BR), Soim (BR)

Kulina's Moustached Tamarin	*Tamarinus kulina*	NE	Brazil

Sauim-de-bigode (BR), Soim (BR)

Geoffroy's Red-bellied Tamarin	*Tamarinus labiatus labiatus*	LC	Bolivia, Brazil, Peru

Tamarino labiado (BO), Leoncito (BO), Sauim (BR), Soim (BR), Pichico de barriga anaranjada (PE), Huapito (PE)

Gray's Red-bellied Tamarin	*Tamarinus labiatus rufiventer*	LC	Brazil

Sauim (BR), Soim (BR)

Thomas's Red-bellied Tamarin	*Tamarinus thomasi*	LC	Brazil

Sauim (BR), Soim (BR)

Black-chinned Emperor Tamarin	*Tamarinus imperator*	LC	Brazil, Peru

Mono bigotudo (BO), Tamarino bigotudo (BO), Tití emperador (BO), Bigodeiro (BR), Pichico emperador (PE), Pichico bigotudo (PE)

Bearded Emperor Tamarin	*Tamarinus subgrisescens*	LC	Bolivia, Brazil, Peru

Mono bigotudo (BO), Tamarino bigotudo (BO), Tití emperador (BO), Bigodeiro (BR), Pichico emperador (PE), Pichico bigotudo (PE)

Mottled-face Tamarin	*Tamarinus inustus*	LC	Brazil, Colombia

Sauim-de-cara-manchada (BR), Soim (BR), Tití diablito (CO), Mico diablo (CO)

Golden-handed Tamarin	*Saguinus midas*	LC	Brazil, French Guiana, Guyana, Suriname

Red-handed Tamarin; Sauim-de-mão-dourada (BR), Sauim-de-mão-vermelha (BR), Suim (BR), Soim (BR), Tamarin à mains dorées (GF), Kusi (SR), Saguwenke (SR), Sagoewintje (SR-NL), Roohandtamarin (SR), Surinaamse Zijde-aap (SR)

Western Black-handed Tamarin	*Saguinus niger*	VU	Brazil

Souim-de-mão-preta-do-oeste (BR), Souim (BR), Choim (BR), Suim (BR), Guaribinha (BR)

Eastern Black-handed Tamarin	*Saguinus ursula*	VU	Brazil

Souim-de-mão-preta-do-leste (BR), Souim (BR), Choim (BR), Suim (BR), Guaribinha (BR)

Pied Tamarin	*Saguinus bicolor*	CR	Brazil

Sauim-de-coleira (BR), Sauim de-duas-cores (BR), Sauim-de-Manaus (BR), Sauim-bicolor (BR)

Martins's Bare-faced Tamarin	*Saguinus martinsi martinsi*	NT	Brazil

Sauim (BR), Souim (BR), Choim (BR), Suim (BR)

NEOTROPICAL PRIMATES

Ochraceous Bare-faced Tamarin	*Saguinus martinsi ochraceus*	NT	Brazil

Sauim (BR), Soim (BR), Choim (BR), Suim (BR)

White-footed Tamarin	*Oedipomidas leucopus*	VU	Colombia

Tití gris (CO)

Cotton-top Tamarin	*Oedipomidas oedipus*	CR	Colombia

Tití cabeciblanco (CO), Tití cabeza de algodón (CO)

Geoffroy's Tamarin	*Oedipomidas geoffroyi*	NT	Colombia, Panama

Tití del Chocó (CO), Tití cabeciblanco (CO), Bichichi (CO), Mono tití panameño (PA)

Golden Lion Tamarin	*Leontopithecus rosalia*	EN	Brazil

Mico-leão-dourado (BR), Saui-piranga (BR), Mico-leão-vermelho (BR)

Golden-headed Lion Tamarin	*Leontopithecus chrysomelas*	EN	Brazil

Mico-leão-de-cara-dourada (BR), Mico-leão-baiano (BR)

Black Lion Tamarin	*Leontopithecus chrysopygus*	EN	Brazil

Mico-leão-preto (BR)

Black-faced Lion Tamarin	*Leontopithecus caissara*	EN	Brazil

Mico-leão-de-cara-preta (BR), Mico-leão-caiçara (BR)

SQUIRREL MONKEYS, GRACILE CAPUCHINS AND ROBUST CAPUCHINS · CEBIDAE · p. 68

Black-crowned Central American Squirrel Monkey	*Saimiri oerstedii oerstedii*	EN	Costa Rica, Panama

Mono tití (CR, PA), Mono tití chiricano (PA), Mono ardilla (PA)

Grey-crowned Central American Squirrel Monkey	*Saimiri oerstedii citrinellus*	EN	Costa Rica

Mono tití (CR)

Humboldt's Squirrel Monkey	*Saimiri cassiquiarensis*	LC	Brazil, Colombia, Venezuela

Macaco-de-cheiro-de-cabeça-branca (BR), Mono ardilla (CO)

Colombian Squirrel Monkey	*Saimiri albigena*	VU	Colombia

Mono ardilla (CO)

Ecuadorian Squirrel Monkey	*Saimiri macrodon*	LC	Brazil, Colombia, Ecuador, Peru

Macaco-de-cheiro-comum (BR), Mono ardilla (CO, EC), Barizo (EC), Fraile (PE), Mono ardilla (PE), Huasa (PE)

Golden-backed Squirrel Monkey	*Saimiri ustus*	NT	Brazil

Macaco-de-cheiro (BR), Macaco-esquilo (BR)

Guianan Squirrel Monkey	*Saimiri sciureus*	LC	Brazil, French Guiana, Guyana, Suriname, Venezuela

Macaco-de-cheiro (BR), Boca-preta (BR), Jurupari (BR), Saïmiri (FG), Monki-monki (SR), Doodshoofdaapje (SR-NL), Doodskosaap (SR-NL)

Collins's Squirrel Monkey	*Saimiri collinsi*	LC	Brazil

Macaco-de-cheiro (BR)

Bolivian Squirrel Monkey	*Saimiri boliviensis boliviensis*	LC	Bolivia, Brazil, Peru

Macaco-de-cheiro (BR), Chichilo (BO), Fraile (PE), Mono ardilla (PE), Huasa (PE), Frailecillo (Peru)

Peruvian Squirrel Monkey	*Saimiri boliviensis peruviensis*	LC	Peru

Fraile (PE), Mono ardilla (PE), Huasa (PE)

| | Black-headed Squirrel Monkey | *Saimiri vanzolinii* | EN | Brazil |

Macaco-de-cheiro-de-cabeça-preta (BR)

| | Northern Black-horned Capuchin | *Sapajus nigritus* | NT | Brazil |

Macaco-prego (BR)

| | Southern Black-horned Capuchin | *Sapajus cucullatus* | NT | Brazil, Argentina |

Macaco-prego (BR), Caí (AR)

| | Hooded Capuchin | *Sapajus cay* | VU | Argentina, Bolivia, Brazil, Paraguay |

Caí (AR, PY), Silbador (BO), Mono martín (BO), Kai (BO), Macaco-prego-do-papo-amarelo (BR), Mono capuchino de Azara (PY)

| | Crested Capuchin | *Sapajus robustus* | EN | Brazil |

Macaco-prego-de-crista (BR)

| | Bearded Capuchin | *Sapajus libidinosus* | NT | Brazil |

Macaco-prego (BR)

| | Yellow-breasted Capuchin | *Sapajus xanthosternos* | CR | Brazil |

Macaco-prego-de-peito amarelo (BR), Pichicau (BR)

| | Blond Capuchin | *Sapajus flavius* | EN | Brazil |

Macaco-prego-dourado (BR), Macaco-prego-galego (BR)

| | Guianan Brown Capuchin | *Sapajus apella apella* | LC | Brazil, Colombia, French Guiana, Guyana, Suriname, Venezuela |

Silbador (BO), Macaco-prego (BR), Capucin Brun (GF), Sajou brun (GF), Keskesi (SR), Bruine Kapucijnaap (SR-NL), Zwarte Capucijneraap (SR-NL)

| | Margarita Island Capuchin | *Sapajus apella margaritae* | CR | Venezuela (Isla de Margarita) |

Mono capuchino pardo (VE)

| | Large-headed Capuchin | *Sapajus macrocephalus* | LC | Bolivia, Brazil, Colombia, Ecuador, Peru |

Macaco-prego (BR), Maicero (CO), Barizo (CO), Machin café (EC), Capuchino negro (EC) Machín negro (PE), Mono negro (PE), Mono capuchino pardo (VE)

| | Peruvian White-fronted Capuchin | *Cebus yuracus* | NT | Brazil, Colombia, Ecuador, Peru |

Caiarara (BR), Capuchino de frente blanca (EC), capuchino blanco del Marañón (EC), Machín blanco (PE), Mono blanco (PE)

| | Shock-headed Capuchin | *Cebus cuscinus* | NT | Bolivia, Peru |

Toranzo (BO), Machín blanco (BO, PE), Mono blanco (PE), Machín frontiblanco (PE)

| | Spix's White-fronted Capuchin | *Cebus unicolor* | VU | Bolivia, Brazil, Peru |

Toranzo (BO), Caiarara (BR), Machín blanco (BO, PE), Mono blanco (PE)

| | Humboldt's White-fronted Capuchin | *Cebus albifrons* | LC | Brazil, Colombia, Venezuela |

Caiarara (BR), Mono cariblanco (CO), Maicero cariblanco (CO), Mono blanco (PE), Machin blanco (PE), Mono capuchino cariblanco (VE)

| | Guianan Weeper Capuchin | *Cebus olivaceus* | LC | Brazil, Guyana, Venezuela |

Caiarara (BR), Caiara (BR), Mono capuchino común (VE)

| | Chestnut Weeper Capuchin | *Cebus castaneus* | LC | Brazil, French Guiana, Guyana, Suriname |

Caiarara (BR), Capucin à tête blanche (GF), Bergi keskesi (SR), Grijze Kapucijnaap (SR-NL)

| | Ka'apor Capuchin | *Cebus kaapori* | CR | Brazil |

Caiarara (BR)

| | Sierra de Perijá White-fronted Capuchin | *Cebus leucocephalus* | VU | Colombia, Venezuela |

Mono cariblanco (CO), Mono capuchino cariblanco (VE)

NEOTROPICAL PRIMATES

Varied White-fronted Capuchin	*Cebus versicolor*	EN	Colombia

Mono cariblanco (CO)

Río Cesar White-fronted Capuchin	*Cebus cesarae*	EN	Colombia

Mono cariblanco (CO)

Santa Marta White-fronted Capuchin	*Cebus malitiosus*	EN	Colombia

Mono cariblanco (CO)

Trinidad White-fronted Capuchin	*Cebus trinitatis*	CR	Trinidad

Weeping Capuchin, Matchin

Ecuadorian White-fronted Capuchin	*Cebus aequatorialis*	CR	Ecuador, Peru

Capuchino ecuatorial (EC), Capuchino blanco ecuatoriano (EC), Machín blanco (PE), Mono blanco (PE), Mono machín de Tumbes (PE), Mono Mico (PE)

Colombian White-faced Capuchin	*Cebus capucinus capucinus*	EN	Panama, Colombia, Ecuador

Maicero capuchino (CO), Mono capuchino (CO, PA), Capuchino de cara blanca (EC), Mono cariblanco (PA)

Gorgona White-faced Capuchin	*Cebus capucinus curtus*	VU	Colombia (Gorgona Is.)

Mono capuchino (CO)

Panamanian White-faced Capuchin	*Cebus imitator*	VU	Costa Rica, Honduras, Nicaragua, Panama (including Coiba Is. & Jicarón Is.)

Mono carita (PA), Mono carilla (PA), Mono carablanca (PA, CR, HN, NI), Mono capuchino (PA, CR, HN, NI), Cariblanco (PA, CR, HN, NI)

NIGHT MONKEYS · AOTIDAE · p. 81

Panamanian Night Monkey	*Aotus zonalis*	NT	Costa Rica (?), Panama, Colombia

Mono nocturno (CO), Mico de noche chocoano (CO), Marteja (CO, PA), Mono de noche (PA), Jujuná (PA)

Lemurine Night Monkey	*Aotus lemurinus*	VU	Colombia, Ecuador, Venezuela

Mono nocturno (CO), Marteja (CO), Mico de noche andino (CO), Mono nocturno lemurino (EC)

Grey-legged Night Monkey	*Aotus griseimembra*	VU	Colombia, Venezuela

Mono nocturno (CO), Marteja (CO), Mico de noche caribeño (CO)

Brumback's Night Monkey	*Aotus brumbacki*	VU	Colombia

Mono nocturno (CO), Marteja (CO), Mico de noche llanero (CO)

Humboldt's Night Monkey	*Aotus trivirgatus*	LC	Brazil, Colombia, Venezuela

Macaco-da-noite-de-pescoço-cinza (BR), Mono nocturno (CO), Marteja (CO), Mono de noche (VE)

Spix's Night Monkey	*Aotus vociferans*	LC	Brazil, Colombia, Ecuador, Peru

Macaco-da-noite-de-pescoço-cinza (BR), Mono nocturno (CO), Marteja (CO), Mico de noche amazónico (CO), Mono nocturno de Spix (EC), Musmuqui (PE), Buri-buri (PE)

Hernández-Camacho's Night Monkey	*Aotus jorgehernandezi*	DD	Colombia

Mono nocturno (CO), Marteja (CO)

Andean Night Monkey	*Aotus miconax*	EN	Peru

Musmuqui (PE), Mono lhuza (PE), Tutamono (PE), Mono nocturno andino (PE), Tutacho (PE)

Ma's Night Monkey	*Aotus nancymai*	VU	Brazil, Colombia, Peru

Macaco-da-noite-de-Ma (BR), Mono-nocturno-de-Ma (BR), Mono nocturno (CO), Marteja (CO), Musmuqui (PE), Mono lechuza (PE)

Black-headed Night Monkey	*Aotus nigriceps*	LC	Bolivia, Brazil, Peru

Mono nocturno (BO), Cuatro ojos (BO), Mono-michi (BO), Macaco-da noite-de-pescoço-vermelho (BR), Musmuqui (PE), Mono lechuza (PE), Mono nocturno cabecinegro (PE)

Azara's Night Monkey — *Aotus azarae azarae* — DD — Argentina, Bolivia, Brazil, Paraguay
Miriquiná (AR), Cuatro ojos (BO), Lechuza (BO), Mono-michi (BO), Mono nocturno (BO, PY), Macaco-da-noite (BR)

Bolivian Night Monkey — *Aotus azarae boliviensis* — DD — Bolivia, Peru
Cuatro ojos (BO), Mono nocturno (BO), Lechuza (BO), Mono lechuza (PE), Mono nocturno de Azara (PE)

Feline Night Monkey — *Aotus azarae infulatus* — LC — Brazil
Macaco-da-noite (BR)

TITI MONKEYS, SAKIS, BEARDED SAKIS AND UACARIS · PITHECIIDAE · p. 86

Río Beni Titi — *Plecturocebus modestus* — EN — Bolivia
Lucachi (BO)

Olalla Brothers' Titi — *Plecturocebus olallae* — CR — Bolivia
Ururo (BO)

White-eared Titi — *Plecturocebus donacophilus* — LC — Bolivia, Brazil
Faca-faca (BO), Luchachi (BO), Zogue-zogue-de-orelha-branca (BR)

Pale Titi — *Plecturocebus pallescens* — LC — Bolivia, Brazil, Paraguay
Ururo (BO), Zogue-zogue-pálido (BR), Mono tití (PY), Tití chaqueño (PY), Tití del Chaco (PY)

Ornate Titi — *Plecturocebus ornatus* — VU — Colombia
Zogui zogui (CO), Mico tocón (CO), Tocón (CO)

Caquetá Titi — *Plecturocebus caquetensis* — CR — Colombia
Zogui zogui (CO), Mico tocón (CO), Tocón (CO), Cotudo (CO), Zocay (CO), Chortantungue (CO)

San Martín Titi — *Plecturocebus oenanthe* — CR — Peru
Titi do Río Mayo (PE), Mono tocón de San Martín (PE), Tocón colorado (PE), Tocón cobrizo (PE)

Coppery Titi — *Plecturocebus cupreus* — LC — Brazil, Peru
Zogue-zogue-acobreado (BR), Guigó-acobreado (BR), Mico tocón (CO), Zogui zogui (CO), Tocón colorado (PE), Tocón cobrizo (PE)

Red-crowned Titi — *Plecturocebus discolor* — LC — Colombia, Ecuador, Peru
Zogui zogui (CO), Cotoncillo rojo (EC), Tití rojizo (EC)

Chestnut-bellied Titi — *Plecturocebus caligatus* — LC — Brazil
Zogue-zogue (BR)

Doubtful Titi — *Plecturocebus dubius* — LC — Brazil
Zogue-zogue-de-Hershkovitz (BR), Guigó-duvidoso (BR)

Stephen Nash's Titi — *Plecturocebus stephennashi* — DD — Brazil
Zogue-zogue-de-Stephen-Nash (BR), Guigó de-Stephen-Nash (BR)

Ashy Titi — *Plecturocebus cinerascens* — LC — Brazil
Zogue-zogue-cinzento (BR), Guigó-cinzento (BR)

Milton's Titi — *Plecturocebus miltoni* — DD — Brazil
Zogue-zogue-de-rabo-de-fogo (BR), Guigó-rabo-de-fogo (BR)

Hoffmanns's Titi — *Plecturocebus hoffmannsi* — LC — Brazil
Zogue-zogue-de-Hoffmanns (BR), Guigó-de-Hoffmanns (BR)

Lake Baptista Titi — *Plecturocebus baptista* — LC — Brazil
Zogue-zogue-do-Lago-Baptista (BR), Guigó-do-Lago-Baptista (BR)

Red-bellied Titi	*Plecturocebus moloch*	LC	Brazil	
Zogue-zogue-vermelho-inchado (BR), Guigó-vermelho-inchado (BR)				
Vieira's Titi	*Plecturocebus vieirai*	CR	Brazil	
Zogue-zogue-de-Vieira (BR), Guigó-de-Vieira (BR)				
Groves's Titi	*Plecturocebus grovesi*	CR	Brazil	
Zogue-zogue-do-Mato-Grosso (BR)				
Brown Titi	*Plecturocebus brunneus*	VU	Bolivia, Brazil	
Zogue-zogue-marrom (BR), Guigó-marrom (BR)				
Prince Bernhard's Titi	*Plecturocebus bernhardi*	LC	Brazil	
Zogue-zogue-do-Príncipe-Bernhard (BR), Guigó-de-Príncipe-Bernhard (BR)				
Toppin's Titi	*Plecturocebus toppini*	LC	Bolivia, Brazil, Peru	
Zogue-zogue-de-Toppin (BR), Tocón colorado (PE)				
Urubamba Brown Titi	*Plecturocebus urubambensis*	LC	Peru	
Tocón de Urubamba (PE)				
Madidi Titi	*Plecturocebus aureipalatii*	LC	Bolivia, Peru	
Luca-luca (BO), Lucachi (BO), Zogue-zogue (BR)				
Medem's Titi	*Cheracebus medemi*	VU	Colombia	
Zogui zogui (CO), Viudita (CO)				
White-collared Titi	*Cheracebus torquatus*	LC	Brazil	
Zogue-zogue-de-colarinho-branco (BR), Guigó-de-colarinho-branco (BR)				
White-chested Titi	*Cheracebus lugens*	LC	Brazil, Colombia, Venezuela	
Zogue-zogue-de-peito-branco (BR), Guigó-de-peito-branco (BR), Zogui zogui (CO), Viudita (CO), Mono viudita (VE)				
Yellow-handed Titi	*Cheracebus lucifer*	LC	Brazil, Colombia, Ecuador, Peru	
Zogue-zogue-de-mão-amarela (BR), Guigó-de-mão-amarela (BR), Zogui zogui (CO), Viudita (CO), Cotoncillo de manos amarillas (EC), Tití de manos amarillas (EC), Tocón negro (PE), Tocón de collar (PE)				
Juruá Collared Titi	*Cheracebus regulus*	LC	Brazil	
Zogue-zogue-do-rio-Juruá (BR), Guigó-de-colar-do-rio-Juruá (BR)				
Aquino's Collared Titi	*Cheracebus aquinoi*	NE	Peru	
Mono tocón de Aquino (PE)				
Black-fronted Titi	*Callicebus nigrifrons*	NT	Brazil	
Guigó-de-frente-preto (BR), Sauá-de-cara-preta (BR), Guigó (BR)				
Masked Titi	*Callicebus personatus*	VU	Brazil	
Guigó-mascarado (BR), Sauá-mascarado (BR)				
Southern Bahian Titi	*Callicebus melanochir*	VU	Brazil	
Guigó-cinza (BR), Guigó-do-sul-da-Bahia (BR)				
Coimbra-Filho's Titi	*Callicebus coimbrai*	EN	Brazil	
Guigó-de-Coimbra-Filho (BR)				
Blond Titi	*Callicebus barbarabrownae*	CR	Brazil	
Guigó-loiro (BR), Guigó-da-Caatinga (BR)				
White-faced Saki	*Pithecia pithecia*	LC	Brazil, French Guiana, Guyana, Suriname	
Parauacu-de-cara-branca (BR), Saki à face pâle (FG), White-faced huruwa (GY), Wanakoe (SR), Witkopsaki (SR-NL), Mono viudo (VE)				

Golden-faced Saki	*Pithecia chrysocephala*	LC	Brazil
Parauacu (BR)			
Hairy Saki	*Pithecia hirsuta*	LC	Brazil, Colombia, Peru
Parauacu (BR), Mono volador (CO), Huapo (CO), Huapo negro (PE)			
Miller's Saki	*Pithecia milleri*	VU	Colombia, Ecuador, Peru
Mono volador (CO), Huapo (CO), Parahuaco (EC), Saki de Miller (EC)			
Geoffroy's Monk Saki	*Pithecia monachus*	LC	Brazil, Peru
Parauacu-monge (BR), Huapo negro (PE), Yana huapo (PE)			
Burnished Saki	*Pithecia inusta*	LC	Brazil, Peru
Parauacu (BR), Huapo negro (PE)			
Cazuza's Saki	*Pithecia cazuzai*	DD	Brazil
Parauacu (BR)			
Equatorial Saki	*Pithecia aequatorialis*	LC	Ecuador, Peru
Parahuaco (EC), Saki ecuatorial (EC), Huapo negro (PE), Huapo Ecuatorial (PE)			
Napo Saki	*Pithecia napensis*	LC	Ecuador, Peru
Parahuaco (EC), Saki del Napo (EC), Huapo negro (PE)			
Isabel's Saki	*Pithecia isabela*	DD	Peru
Huapo (PE)			
Buffy Saki	*Pithecia albicans*	LC	Brazil
Parauacu-castanho-amarelado (BR)			
Gray's Bald-faced Saki	*Pithecia irrorata*	DD	Bolivia, Brazil, Peru
Parahuaco (BO), Parauacú-de-Gray (BR), Macaco-parauacú (BR), Huapo de Gray (PE)			
Vanzolini's Bald-faced Saki	*Pithecia vanzolinii*	DD	Brazil
Parauacu-de-Vanzolini (BR)			
Mittermeier's Tapajós Saki	*Pithecia mittermeieri*	VU	Brazil
Parauacu (BR)			
Pissinatti's Bald-faced Saki	*Pithecia pissinattii*	DD	Brazil
Parauacu (BR)			
Red-nosed Bearded Saki	*Chiropotes albinasus*	VU	Brazil
White-nosed Saki; Cuxiú-de-nariz-vermelho (BR)			
Black Saki	*Chiropotes satanas*	EN	Brazil
Cuxiú-preto (BR)			
Rio Negro Bearded Saki	*Chiropotes chiropotes*	LC	Brazil, Venezuela
Cuxiú-do-Rio-Negro (BR), Mono barbudo (VE)			
Uta Hick's Bearded Saki	*Chiropotes utahicki*	VU	Brazil
Cuxiú-de-Uta-Hick (BR)			
Guianan Bearded Saki	*Chiropotes sagulatus*	LC	Brazil, French Guiana, Guyana, Suriname
Red-backed Bearded Saki, Cuxiú (BR), Saki satan (FG), Saki capucin (FG), Bisa (SR), Barba-man (SR), Wiché (SR), Baardsaki (SR-NL)			
White Bald Uacari	*Cacajao calvus*	LC	Brazil
Uacari-branco (BR)			

NEOTROPICAL PRIMATES

Novaes's Bald Uacari	*Cacajao novaesi*	VU	Brazil

Uacari-de-Novaes (BR)

Red Bald Uacari	*Cacajao rubicundus*	LC	Brazil, Colombia (?)

Uacari-vermelho (BR)

Ucayali Bald Uacari	*Cacajao ucayalii*	VU	Brazil, Peru

Uacari-do-Rio-Ucayali (BR), Huapo rojo (PE), Huapo colorado (PE), Mono inglés (PE), Puca huapo (PE)

Kanamari Bald Uacari	*Cacajao amuna*	NE	Brazil

Uacari-dos-Kanamari (BR)

Humboldt's Black-headed Uacari	*Cacajao melanocephalus*	LC	Brazil, Venezuela

Uacari-preto (BR), Macaco-Bicó (BR), Acarí (BR), Acarí-bicó (BR), Colimocho (CO), Ichacha (CO), Mono chucuto (VE)

Neblina Black-headed Uacari	*Cacajao hosomi*	VU	Brazil, Colombia, Venezuela

Uacari-preto-de-Neblina (BR), Colimocho (CO), Mono chucuto (VE)

Ayres's Black-headed Uacari	*Cacajao ayresi*	LC	Brazil

Uacari-preto- de-Araca (BR)

HOWLER MONKEYS, SPIDER MONKEYS, WOOLLY MONKEYS AND MURIQUIS · ATELIDAE · p. 111

Colombian Red Howler	*Alouatta seniculus seniculus*	LC	Brazil, Colombia, Ecuador, Peru, Venezuela

Guariba-vermelha (BR), Aullador colorado (CO), Aullador rojo (CO, EC), Coto rojo (EC), Aullador colorado (CO), Araguato (PE), Coto (PE), Coto mono (PE), Aullador rojizo (PE), Keníri (PE), Yaniri (PE), Mono araguato (VE)

Trinidad Red Howler	*Alouatta seniculus insulanus*	NE	Trinidad

Guianan Red Howler Monkey, Howler; Macaque rouge

Juruá Red Howler	*Alouatta juara*	LC	Brazil, Peru

Guariba-vermelha-do-Juruá (BR)

Purus Red Howler	*Alouatta puruensis*	VU	Brazil, Peru

Guariba-vermelha-do-Purus (BR)

Ursine Red Howler	*Alouatta arctoidea*	LC	Venezuela

Mono araguato (VE)

Guianan Red Howler	*Alouatta macconnelli*	LC	Brazil, French Guiana, Guyana, Suriname, Venezuela

Guariba-vermelha-dos-Guianas (BR), Hurleur roux (GF), Baboen (SR), Rode brulaap (SR-NL), Mono araguato (VE)

Bolivian Red Howler	*Alouatta sara*	NT	Bolivia, Brazil, Peru

Manechi colorado (BO), Guariba-vermelha-de-Bolivia (BR), Coto (PE), Mono aullador (PE)

Amazon Black Howler	*Alouatta nigerrima*	LC	Brazil

Guariba-preta-da-Amazônia (BR)

Red-Handed Howler	*Alouatta belzebul*	VU	Brazil

Guariba-de-mãos-ruivas (BR), Guariba-de-mãos-vermelhas (BR), Guariba-preta (BR), Bugio-de-mãos-ruivas (BR), Guariba (BR), Górgo (BR)

Spix's Howler	*Alouatta discolor*	VU	Brazil

Guariba-preta (BR)

Maranhão Red-handed Howler	*Alouatta ululata*	EN	Brazil

Guariba-preta-de-Maranhão (BR), Guariba-da-Caatinga (BR)

Brown Howler	*Alouatta guariba*	VU	Argentina, Brazil

Aullador rojo (Ar), Bugio (BR), Bugio-ruivo (BR), Barbado (BR)

Black and Gold Howler	*Alouatta caraya*	NT	Argentina, Bolivia, Brazil, Paraguay, Uruguay

Mono aullador negro y dorado (AR, PY), Carayá negro y dorado (AR, PY), Carayá (AR, PY), Manechi negro (BO), Bugio-preto (BR), Bugio-preto-e-dorado (BR)

Golden-mantled Howler	*Alouatta palliata palliata*	EN	Costa Rica, Guatemala, Honduras, Nicaragua, Panama (?)

Mono aullador (CO), Mono Congo (CR), Aullador de manto dorado (EC), Mono Zaraguate (GT), Mono negro (PA)

Azuero Peninsula Howler	*Alouatta palliata trabeata*	EN	Panama

Mono gun gun (PA), Mono kun kun (PA)

Coiba Island Howler	*Alouatta palliata coibensis*	EN	Panama (Coiba Is, Jicarón Is)

Mono aullador de Coiba (PA)

Ecuadorian Mantled Howler	*Alouatta palliata aequatorialis*	VU	Colombia, Ecuador, Panama, Peru

Aullador Negro (CO, EC), Mono coto de Tumbes (PE), Mono aullador (PA)

Mexican Mantled Howler	*Alouatta palliata mexicana*	EN	Mexico

Zaraguate (MX), Saraguato café (MX), Mono aullador de manto (MX)

Central American Black Howler	*Alouatta pigra*	EN	Belize, Guatemala, Mexico

Baboon (BZ), Mono Zaraguate (GT), Saraguato negro (MX), Mono aullador negro (MX)

Geoffroy's Spider Monkey	*Ateles geoffroyi geoffroyi*	CR	Nicaragua

Mono colorado (CR)

Azuero Spider Monkey	*Ateles geoffroyi azuerensis*	CR	Panama

Mono charro (PA), Mono charao (PA), Mani larga (PA)

Black-browed Spider Monkey	*Ateles geoffroyi frontatus*	VU	Costa Rica, Nicaragua

Mono colorado (CR)

Hooded Spider Monkey	*Ateles geoffroyi grisescens*	DD	Panama, Colombia (?)

Mono araña gris del Darién (PA)

Ornate Spider Monkey	*Ateles geoffroyi ornatus*	VU	Costa Rica, Nicaragua

Mono colorado (CR)

Mexican Spider Monkey	*Ateles geoffroyi vellerosus*	EN	Belize, El Salvador, Guatemala, Honduras, Mexico

Monkey (BZ), Mono colorado (GT), Mono araña (MX)

Colombian Black Spider Monkey	*Ateles rufiventris*	VU	Colombia, Panama

Marimonda (CO), Marimonda chocoana (CO), Mono araña (CO, PA), Mono araña negro del Darién (PA) Mono negro (PA), Mono yerre (PA)

Ecuadorian Spider Monkey	*Ateles fusciceps*	CR	Ecuador

Mono araña de cabeza marrón (EC)

Black Spider Monkey	*Ateles chamek*	EN	Bolivia, Brazil, Peru

Marimono (BO), Mono araña negro (BO, PE), Macaco-aranha-de-cara-preta (BR), Coatá-de-cara-preta (BR), Mono araña (BO, PE), Maquisapa (PE), Covéro (PE), Oshéto (PE)

Red-faced Black Spider Monkey	*Ateles paniscus*	VU	Brazil, French Guiana, Guyana, Suriname

Macaco-aranha-de-cara-vermelha (BR), Coatá-de-cara-vermelha (BR), Atèle noir (GF), Kwatta (SR), Zwarte Spinaap (SR-NL)

White-whiskered Spider Monkey	*Ateles marginatus*	EN	Brazil

Macaco-aranha-de-cara-branca (BR), Coatá-de-cara-branca (BR), Macaco-aranha-de-testa-branca (BR)

NEOTROPICAL PRIMATES

White-bellied Spider Monkey — *Ateles belzebuth* — EN — Brazil, Colombia, Ecuador, Peru, Venezuela
Macaco-aranha-de-barriga-branca (BR), Mono araña (CO, PE), Marimba (CO), Marimonda amazónica (CO), Mono araña de vientre amarillo (EC), Maquisapa (PE), Mono araña grisáceo (PE), Koshíri (PE), Iempari (PE), Mono araña Común (VE)

Variegated Spider Monkey — *Ateles hybridus* — CR — Colombia, Venezuela
Mono araña café (CO), Marimonda del Magdalena (CO), Choibo (CO), Mono araña Común (VE)

Humboldt's Woolly Monkey — *Lagothrix lagothricha lagothricha* — VU — Brazil, Colombia, Ecuador, Peru
Macaco-barrigudo (BR), Churuco (CO), Mono lanudo cenizo (EC), Chorongo cenizo (EC), Mono lanudo (EC, PE), Choro (PE)

Colombian Woolly Monkey — *Lagothrix lagothricha lugens* — CR — Colombia
Churuco (CO), Mono lanudo (CO)

Grey Woolly Monkey — *Lagothrix lagothricha cana* — EN — Brazil
Macaco-barrigudo (BR), Choro (PE), Mono lanudo (PE)

Poeppig's Woolly Monkey — *Lagothrix lagothricha poeppigii* — EN — Brazil, Ecuador, Peru
Macaco-barrigudo (BR), Mono lanudo rojizo (EC), Chorongo rojizo (EC), Choro (PE), Mono lanudo (PE), Mono choro común (PE)

Peruvian Woolly Monkey — *Lagothrix lagothricha tschudii* — DD — Bolivia, Peru
Mono barrigudo (BO), Marimondo de frío (BO), Mono rosillo (BO), Mono lanudo (BO, PE), Choro (PE), Mono choro (PE)

Peruvian Yellow-tailed Woolly Monkey — *Lagothrix flavicauda* — CR — Peru
Choro de cola amarilla (PE), Mono choro de cola amarilla (PE), Tupa (PE), Choba (PE)

Southern Muriqui — *Brachyteles arachnoides* — CR — Brazil
Muriqui-do-sul (BR)

Northern Muriqui — *Brachyteles hypoxanthus* — CR — Brazil
Muriqui-do-norte (BR), Mono carvoeiro (BR)

CARIBBEAN AFRICAN MONKEYS (INTRODUCED) · CERCOPITHECIDAE · p. 125

Green Monkey — *Chlorocebus sabaeus* — LC — Barbados, St. Kitts and Nevis, St. Lucia, Sint Eustatius

Vervet Monkey — *Chlorocebus pygerythrus* — LC — Saint-Martin/Sint Maarten & Anguilla

Mona Monkey — *Cercopithecus mona* — NT — Grenada
Mona Guenon (GD)

INDEX

A

Alouatta arctoidea 112
Alouatta belzebul 113
Alouatta caraya 115
Alouatta discolor 114
Alouatta guariba 115
Alouatta juara 111
Alouatta macconnelli 112
Alouatta nigerrima 113
Alouatta palliata aequatorialis 117
Alouatta palliata coibensis 116
Alouatta palliata mexicana 117
Alouatta palliata palliata 116
Alouatta palliata trabeata 116
Alouatta pigra 117
Alouatta puruensis 112
Alouatta sara 113
Alouatta seniculus insulanus 111
Alouatta seniculus seniculus 111
Alouatta ululata 114
Amazon Black Howler 113
Andean Night Monkey 83
Andean Saddle-back Tamarin 56
Aotus azarae azarae 84
Aotus azarae boliviensis 85
Aotus azarae infulatus 85
Aotus brumbacki 82
Aotus griseimembra 81
Aotus jorgehernandezi 83
Aotus lemurinus 81
Aotus miconax 83
Aotus nancymai 83
Aotus nigriceps 84
Aotus trivirgatus 82
Aotus vociferans 82
Aotus zonalis 81
Aquino's Collared Titi 96
Ashy Titi 90
Ateles belzebuth 121
Ateles chamek 120
Ateles fusciceps 120
Ateles geoffroyi azuerensis 118
Ateles geoffroyi frontatus 118
Ateles geoffroyi geoffroyi 118
Ateles geoffroyi grisescens 119
Ateles geoffroyi ornatus 119
Ateles geoffroyi vellerosus 119
Ateles hybridus 122
Ateles marginatus 121
Ateles paniscus 121
Ateles rufiventris 120
Ávila-Pires's Saddle-back Tamarin 58
Ayres's Black-headed Uacari 110
Azara's Night Monkey 84
Azuero Peninsula Howler 116
Azuero Spider Monkey 118

B

Bearded Capuchin 73
Bearded Emperor Tamarin 62

Black and Gold Howler 115
Black Lion Tamarin 67
Black Saki 107
Black Spider Monkey 120
Black-and-white Tassel-ear Marmoset 48
Black-browed Spider Monkey 118
Black-chinned Emperor Tamarin 62
Black-crowned Central American Squirrel Monkey 68
Black-crowned Dwarf Marmoset 46
Black-faced Lion Tamarin 67
Black-fronted Titi 96
Black-headed Marmoset 50
Black-headed Night Monkey 84
Black-headed Squirrel Monkey 71
Black-tailed Marmoset 51
Black-tufted-ear Marmoset 52
Blond Capuchin 74
Blond Titi 98
Bolivian Night Monkey 85
Bolivian Red Howler 113
Bolivian Squirrel Monkey 70
Brachyteles arachnoides 124
Brachyteles hypoxanthus 124
Brown Howler 115
Brown Titi 92
Brumback's Night Monkey 82
Buffy Saki 104
Buffy-headed Marmoset 53
Buffy-tufted-ear Marmoset 54
Burnished Saki 101

C

Cacajao amuna 109
Cacajao ayresi 110
Cacajao calvus 108
Cacajao hosomi 110
Cacajao melanocephalus 110
Cacajao novaesi 108
Cacajao rubicundus 109
Cacajao ucayalii 109
Callibella humilis 46
Callicebus barbarabrownae 98
Callicebus coimbrai 98
Callicebus melanochir 97
Callicebus nigrifrons 96
Callicebus personatus 97
Callimico goeldii 52
Callithrix aurita 54
Callithrix flaviceps 53
Callithrix geoffroyi 53
Callithrix jacchus 53
Callithrix kuhlii 53
Callithrix penicillata 52
Caquetá Titi 87
Cazuza's Saki 102
Cebuella niveiventris 46
Cebuella pygmaea 46
Cebus aequatorialis 79

Cebus albifrons 76
Cebus capucinus capucinus 80
Cebus capucinus curtus 80
Cebus castaneus 77
Cebus cesarae 78
Cebus cuscinus 75
Cebus imitator 80
Cebus kaapori 77
Cebus leucocephalus 77
Cebus malitiosus 79
Cebus olivaceus 76
Cebus trinitatis 79
Cebus unicolor 76
Cebus versicolor 78
Cebus yuracus 75
Central American Black Howler 117
Cercopithecus mona 125
Cheracebus aquinoi 96
Cheracebus lucifer 95
Cheracebus lugens 95
Cheracebus medemi 94
Cheracebus regulus 95
Cheracebus torquatus 94
Chestnut Weeper Capuchin 77
Chestnut-bellied Titi 89
Chiropotes albinasus 106
Chiropotes chiropotes 107
Chiropotes sagulatus 108
Chiropotes satanas 107
Chiropotes utahicki 107
Chlorocebus pygerythrus 125
Chlorocebus sabaeus 125
Coiba Island Howler 116
Coimbra-Filho's Titi 98
Collins's Squirrel Monkey 70
Colombian Black Spider Monkey 120
Colombian Red Howler 111
Colombian Squirrel Monkey 69
Colombian White-faced Capuchin 80
Colombian Woolly Monkey 122
Coppery Titi 88
Cotton-top Tamarin 65
Crested Capuchin 73
Cruz Lima's Saddle-back Tamarin 57

D
Doubtful Titi 89

E
Eastern Black-handed Tamarin 63
Ecuadorian Mantled Howler 117
Ecuadorian Spider Monkey 120
Ecuadorian Squirrel Monkey 69
Ecuadorian White-fronted Capuchin 79
Equatorial Saki 102

F
Feline Night Monkey 85

G
Geoffroy's Monk Saki 101
Geoffroy's Red-bellied Tamarin 61
Geoffroy's Saddle-back Tamarin 57

Geoffroy's Spider Monkey 118
Geoffroy's Tamarin 65
Goeldi's Monkey 52
Golden Lion Tamarin 66
Golden-backed Squirrel Monkey 69
Golden-faced Saki 99
Golden-handed Tamarin 63
Golden-headed Lion Tamarin 66
Golden-mantled Howler 116
Golden-mantled Saddle-back Tamarin 55
Golden-white Bare-ear Marmoset 47
Golden-white Tassel-ear Marmoset 49
Gorgona White-faced Capuchin 80
Graells's Black-mantled Tamarin 54
Gray's Bald-faced Saki 104
Gray's Red-bellied Tamarin 61
Green Monkey 125
Grey Woolly Monkey 123
Grey-crowned Central American Squirrel Monkey 68
Grey-fronted Saddle-back Tamarin 58
Grey-legged Night Monkey 81
Groves's Titi 92
Guianan Bearded Saki 108
Guianan Brown Capuchin 74
Guianan Red Howler 112
Guianan Squirrel Monkey 70
Guianan Weeper Capuchin 76

H
Hairy Saki 100
Hernández-Camacho's Black-mantled Tamarin 55
Hernández-Camacho's Night Monkey 83
Hershkovitz's Saddle-back Tamarin 58
Hoffmanns's Titi 91
Hooded Capuchin 72
Hooded Spider Monkey 119
Humboldt's Black-headed Uacari 110
Humboldt's Night Monkey 82
Humboldt's Squirrel Monkey 68
Humboldt's White-fronted Capuchin 76
Humboldt's Woolly Monkey 122

I
Illiger's Saddle-back Tamarin 56
Isabel's Saki 103

J
Juruá Collared Titi 95
Juruá Red Howler 111

K
Ka'apor Capuchin 77
Kanamari Bald Uacari 109
Kulina's Moustached Tamarin 60

L
Lagothrix flavicauda 124
Lagothrix lagothricha cana 123
Lagothrix lagothricha lagothricha 122
Lagothrix lagothricha lugens 122
Lagothrix lagothricha poeppigii 123

Lagothrix lagothricha tschudii 123
Lake Baptista Titi 91
Large-headed Capuchin 75
Lemurine Night Monkey 81
Leontocebus avilapiresi 58
Leontocebus cruzlimai 57
Leontocebus fuscicollis 57
Leontocebus fuscus 55
Leontocebus illigeri 56
Leontocebus lagonotus 56
Leontocebus leucogenys 56
Leontocebus melanoleucus 59
Leontocebus mura 58
Leontocebus nigricollis graellsi 54
Leontocebus nigricollis hernandezi 55
Leontocebus nigricollis nigricollis 54
Leontocebus nigrifrons 57
Leontocebus primitivus 58
Leontocebus tripartitus 55
Leontocebus weddelli 59
Leontopithecus caissara 67
Leontopithecus chrysomelas 66
Leontopithecus chrysopygus 67
Leontopithecus rosalia 66
Lesson's Saddle-back Tamarin 55

M
Ma's Night Monkey 83
Madidi Titi 94
Maranhão Red-handed Howler 114
Marca's Marmoset 50
Margarita Island Capuchin 74
Martins's Bare-faced Tamarin 64
Masked Titi 97
Maués Marmoset 49
Medem's Titi 94
Mexican Mantled Howler 117
Mexican Spider Monkey 119
Mico acariensis 50
Mico argentatus 47
Mico chrysoleucos 49
Mico emiliae 47
Mico humeralifer 48
Mico intermedius 51
Mico leucippe 47
Mico marcai 50
Mico mauesi 49
Mico melanurus 51
Mico munduruku 48
Mico nigriceps 50
Mico rondoni 51
Mico saterei 49
Mico schneideri 48
Miller's Saki 100
Milton's Titi 90
Mittermeier's Tapajós Saki 105
Mona Monkey 125
Mottled-face Tamarin 62
Munduruku Marmoset 48

N
Napo Saki 103
Neblina Black-headed Uacari 110

Northern Black-horned Capuchin 72
Northern Muriqui 124
Northern Pygmy Marmoset 46
Novaes's Bald Uacari 108

O
Ochraceous Bare-faced Tamarin 64
Oedipomidas geoffroyi 65
Oedipomidas leucopus 65
Oedipomidas oedipus 65
Olalla Brothers' Titi 86
Ornate Spider Monkey 119
Ornate Titi 87

P
Pale Titi 87
Panamanian Night Monkey 81
Panamanian White-faced Capuchin 80
Peruvian Squirrel Monkey 71
Peruvian White-fronted Capuchin 75
Peruvian Woolly Monkey 123
Peruvian Yellow-tailed Woolly Monkey 124
Pied Tamarin 64
Pissinatti's Bald-faced Saki 106
Pithecia aequatorialis 102
Pithecia albicans 104
Pithecia cazuzai 102
Pithecia chrysocephala 99
Pithecia hirsuta 100
Pithecia inusta 101
Pithecia irrorata 104
Pithecia isabela 103
Pithecia milleri 100
Pithecia mittermeieri 105
Pithecia monachus 101
Pithecia napensis 103
Pithecia pissinattii 106
Pithecia pithecia 99
Pithecia vanzolinii 105
Plecturocebus aureipalatii 94
Plecturocebus baptista 91
Plecturocebus bernhardi 93
Plecturocebus brunneus 92
Plecturocebus caligatus 89
Plecturocebus caquetensis 87
Plecturocebus cinerascens 90
Plecturocebus cupreus 88
Plecturocebus discolor 89
Plecturocebus donacophilus 86
Plecturocebus dubius 89
Plecturocebus grovesi 92
Plecturocebus hoffmannsi 91
Plecturocebus miltoni 90
Plecturocebus modestus 86
Plecturocebus moloch 91
Plecturocebus oenanthe 88
Plecturocebus olallae 86
Plecturocebus ornatus 87
Plecturocebus pallescens 87
Plecturocebus stephennashi 90
Plecturocebus toppini 93
Plecturocebus urubambensis 93
Plecturocebus vieirai 92

NEOTROPICAL PRIMATES

Poeppig's Woolly Monkey   123
Prince Bernhard's Titi   93
Purus Red Howler   112

R

Red Bald Uacari   109
Red-bellied Titi   91
Red-capped Moustached Tamarin   60
Red-crowned Titi   89
Red-faced Black Spider Monkey   121
Red-handed Howler   113
Red-mantled Saddle-back Tamarin   56
Red-nosed Bearded Saki   106
Rio Acarí Marmoset   50
Rio Aripuanã Marmoset   51
Río Beni Titi   86
Río Cesar White-fronted Capuchin   78
Rio Negro Bearded Saki   107
Rondon's Marmoset   51

S

Saguinus bicolor   64
Saguinus martinsi martinsi   64
Saguinus martinsi ochraceus   64
Saguinus midas   63
Saguinus niger   63
Saguinus ursula   63
Saimiri albigena   69
Saimiri boliviensis boliviensis   70
Saimiri boliviensis peruviensis   71
Saimiri cassiquiarensis   68
Saimiri collinsi   70
Saimiri macrodon   69
Saimiri oerstedii citrinellus   68
Saimiri oerstedii oerstedii   68
Saimiri sciureus   70
Saimiri ustus   69
Saimiri vanzolinii   71
San Martín Titi   88
Santa Marta White-fronted Capuchin   79
Sapajus apella apella   74
Sapajus apella margaritae   74
Sapajus cay   72
Sapajus cucullatus   72
Sapajus flavius   74
Sapajus libidinosus   73
Sapajus macrocephalus   75
Sapajus nigritus   72
Sapajus robustus   73
Sapajus xanthosternos   73
Sateré Marmoset   49
Schneider's Marmoset   48
Shock-headed Capuchin   75
Sierra de Perijá White-fronted Capuchin   77
Silvery Marmoset   47
Snethlage's Marmoset   47
Southern Bahian Titi   97
Southern Black-horned Capuchin   72
Southern Muriqui   124
Southern Pygmy Marmoset   46
Spix's Black-mantled Tamarin   54
Spix's Howler   114
Spix's Moustached Tamarin   59
Spix's Night Monkey   82
Spix's Saddle-back Tamarin   57
Spix's White-fronted Capuchin   76
Stephen Nash's Titi   90

T

Tamarinus imperator   62
Tamarinus inustus   62
Tamarinus kulina   60
Tamarinus labiatus labiatus   61
Tamarinus labiatus rufiventer   61
Tamarinus mystax   59
Tamarinus pileatus pileatus   60
Tamarinus pileatus pluto   60
Tamarinus subgrisescens   62
Tamarinus thomasi   61
Thomas's Red-bellied Tamarin   61
Toppin's Titi   93
Trinidad Red Howler   111
Trinidad White-fronted Capuchin   79

U

Ucayali Bald Uacari   109
Ursine Red Howler   112
Urubamba Brown Titi   93
Uta Hick's Bearded Saki   107

V

Vanzolini's Bald-faced Saki   105
Varied White-fronted Capuchin   78
Variegated Spider Monkey   122
Vervet Monkey   125
Vieira's Titi   92

W

Weddell's Saddle-back Tamarin   59
Western Black-handed Tamarin   63
White Bald Uacari   108
White Saddle-back Tamarin   59
White-bellied Spider Monkey   121
White-chested Titi   95
White-collared Titi   94
White-eared Titi   86
White-faced Marmoset   53
White-faced Saki   99
White-footed Tamarin   65
White-rumped Moustached Tamarin   60
White-tufted-ear Marmoset   52
White-whiskered Spider Monkey   121
Wied's Black-tufted-ear Marmoset   53

Y

Yellow-breasted Capuchin   73
Yellow-handed Titi   95